EPD Congress 2016

TMS2016
145th Annual Meeting & Exhibition

FEBRUARY 14-18 DOWNTOWN NASHVILLE, TENNESSEE MUSIC CITY CENTER

New proceedings volumes from the TMS2016 Annual Meeting:

- 7th International Symposium on High-Temperature Metallurgical Processing
- CFD Modeling and Simulation in Materials Processing 2016
- Characterization of Minerals, Metals, and Materials 2016
- Energy Technology 2016: Carbon Dioxide Management and Other Technologies
- EPD Congress 2016
- Light Metals 2016
- Magnesium Technology 2016
- Rare Metal Technology 2016
- REWAS 2016
- Shape Casting: 6th International Symposium
- TMS 2016 Supplemental Proceedings

2016 EPD Congress

Proceedings of symposia sponsored by
the Extraction & Processing Division (EPD) of
The Minerals, Metals & Materials Society (TMS)

held during

TMS2016
145th Annual Meeting & Exhibition

FEBRUARY 14-18 DOWNTOWN NASHVILLE,
TENNESSEE MUSIC CITY CENTER

Edited by:
**Antoine Allanore, Laura Bartlett, Cong Wang,
Lifeng Zhang, and Jonghyun Lee**

Editors
Antoine Allanore
Laura Bartlett
Cong Wang
Lifeng Zhang
Jonghyun Lee

ISBN 978-3-319-48621-5
ISBN 978-3-319-48111-1 (eBook)
DOI 10.1007/978-3-319-48111-1

Chemistry and Materials Science: Monograph

Published by Springer International Publishers, Switzerland, 2016
Reprint of the original edition published by John Wiley & Sons, Inc., 2016, 978-3-319-48621-5

Printed on acid-free paper

This Springer imprint is published by Springer Nature
The registered company is Springer International Publishing AG
The registered company address is: Gewerbestrasse 11, 6330 Cham, Switzerland

TABLE OF CONTENTS
EPD Congress 2016

Materials Processing Fundamentals

Casting and Solidification Processes

Non-Ferrous Extractive Metallurgy

Iron and Steelmaking - Thermodynamic, Reduction and Physical Metallurgy

Forming, Joining, Sensing: Devices and Applications

Poster Session

Advanced Characterization Techniques for Quantifying and Modeling Deformation

Session IV

EPD COUNCIL 2015–2016

Gerardo Alvear
Pyrometallurgy Cmte Chair
Business Development Manager
Glendora Technology
Brisbane, Queensland, Australia

Randolph Kirchain
Recycling & Environmental Technologies Cmte Chair
Principal Research Scientist
Massachusetts Institute of Technology
Cambridge, MA, USA

Elsa Olivetti
Content Development & Dissemination Cmte Rep
Research Scientist
Massachusetts Institute of Technology
Cambridge, MA, USA

Adam Powell, IV
Education Cmte Rep
Co-Founder and CTO
Metal Oxygen Separation Technologies Inc.
Natick, MA, USA

Soobhankar Pati
Materials & Society Cmte Rep
Assistant Professor
Indian Institute of Technology
Bhubaneswar, India

Robert Hyers
Membership & Student Development Cmte Rep
Chief Technology Officer
Boston Electrometallurgical Corporation
Woburn, MA, USA

John Carpenter
Program Cmte Rep
Technical Staff Member
Los Alamos National Laboratory
Los Alamos, NM, USA

Boyd Davis

EPD HONORS AND AWARDS 2016

Distinguished Lecturer Award
Markus Reuter
Helmholtz Institute Freiberg for Resource Technology

Science Award
"Influence of Lattice Defects on the Grain Growth Kinetics of Nanocrystalline Fluorite"
Metallurgical and Materials Transactions A, January 2014

Mahmoud Abdellatief
Elettra-Sincrotrone Trieste

Andrea Lausi
Elettra-Sincrotrone Trieste

Jasper R. Plaisier
Elettra-Sincrotrone Trieste

Paolo Scardi
University of Trento

"On the Development of Thermo-Kinetic Eh-pH Diagrams"
Metallurgical and Materials Transactions B, December 2012

Subrata Roy
New Gold

Hamidreza Zebardast
Acuren

Edouard Asselin
University of British Columbia

Technology Award
"A Study on the Effect of Pulse Electrodeposition Parameters on the Morphology of Pure Tin Coatings"
Metallurgical and Materials Transactions A, September 2014

Ashutosh Sharma
Indian Institute of Technology Kharagpur

Sumit Bhattacharya
Northwestern University

Siddhartha Das
Indian Institute of Technology Kharagpur

Karabi Das
Indian Institute of Technology Kharagpur

EPD Congress 2016

MATERIALS PROCESSING FUNDAMENTALS

2016 EPD Congress

SYMPOSIUM:
MATERIALS PROCESSING FUNDAMENTALS

Casting and Solidification Processes

Jonghyun Lee,
University of Massachusetts

Cong Wang,
Northeastern University

EPD Congress 2016
Edited by: Antoine Allanore, Laura Bartlett, Cong Wang, Lifeng Zhang, and Jonghyun Lee
TMS (The Minerals, Metals & Materials Society), 2016

INFLUENCE OF SCALE FORMATION ON COPPER ENRICHMENT BEHAVIOUR IN CONTINUOUSLY CAST SLAB

Nan Wang, Shan Yu, Xiaoao Li, Jianhua Xin, Guanghao Chen, Min Chen, Cuihuan Huang

School of Materials and Metallurgy, Northeastern University, Shenyang, China

Keywords: Copper enrichment; Scale formation; Continuously cast slab; Initial slab surface temperature; Nickel content

Abstract

Copper enrichment behavior in continuously cast slab induced by scale formation after continuous cooling in H_2O vapor atmosphere and consecutively in air has been experimentally investigated by combining with SEM-EDS analysis, for the purpose to examine the effects of initial slab surface temperature and contents of copper and nickel. For Slab 1 with copper content of 0.1 percent, no visible copper enrichment is found at scale-substrate interface irrespective of the various scale microstructures formed at different initial slab surface temperatures. For Slab 2 with higher contents of copper and nickel, copper-rich phase accumulates and stays at scale-substrate interface under the condition of lower initial slab surface temperature; while Cu-rich phase in the porous scale formed at higher initial slab surface temperature is noted with simultaneous Ni-enrichment appearing in front of the steel-scale interface. Both a porous scale and high nickel content work together to provide pores for copper-rich phase migration away from interface and occlude within the scale layer.

Introduction

The use of scrap in steelmaking, especially in electric arc steelmaking, induces inevitable residual elements problem. The residual elements such as copper, tin and arsenic whose oxidation potential are less than iron, are retained in the steel because they are difficult to be removed during steelmaking process. During subsequent processing operations such as continuous casting and hot rolling, residual elements tend to solidification segregation, concentration at the steel-scale interface and penetration along grain boundary, which promote cracking known as hot shortness or hair cracks [1]. In the past, enrichment of residual elements during hot rolling and its influence on surface quality of hot rolled plate have been extensively studied [2-6].

On the other hand, significant residual enrichment has also been detected at the oxide scale/matrix interface in continuously cast slab. This has been related to the rapid oxidation of strand surface with heat release to environment in the secondary cooling zone. The past study results on continuously cast products showed that the hot shortness occurrence was mainly in connection with an oxide layer with clear Cu enrichment at the scale-steel interface. Thus, the residual enrichment may depend on the several factors including scale formation, occlusion of the residual elements into the scale and diffusion of these elements away from the interface into the steel [7-9]. In the current work, an experimental was conducted to investigate the copper enrichment induced by scale formation in continuously cast slab, mainly the effects of initial slab surface temperature in secondary cooling zone and contents of copper and nickel in steel.

Experimental

Two types of Cu-bearing continuous cast slabs, namely Slab 1 and Slab 2, with the chemical compositions listed in Table 1 were used. The copper contents in Slab 1 and Slab 2 were 0.10 and 0.52 mass percent respectively. To evaluate the effect of Ni content on copper enrichment, a relatively higher content of nickel (0.422 mass percent) in Slab 2 compared with Slab 1 (0.040% Ni) was maintained. The slab sample preparation is similar to that in the previous work [10]. To ensure that the oxidizing surface of slab sample is the chilled layer composed of fine equiaxed grain, the upper layer of commercial continuously cast slab was first took down and cut into rectangular samples with dimensions of 10 mm ×10 mm ×5 mm. Then the original scale and mold flux adhered to the sample surface were removed using SiC abrasion paper and cleaned ultrasonically in ethanol.

In the laboratory experiment, the prepared slab sample was firstly heated to the desired temperature (1100℃/1300℃) in a horizontal tube furnace and held at the desired temperature for five minutes in argon gas flow. Once the sample surface temperature reached the desired initial temperature, the samples with different initial surface temperatures were then rapidly taken out of the heating furnace and put into another horizontal atmosphere-controlled furnace to cool in H_2O vapor atmosphere to about 800℃. Finally, the sample was removed from the atmosphere-controlled furnace to cool to room temperature in ambient air. The whole experiment process was designed to investigate scale formation and induced copper enrichment during the simulated secondary and air cooling zones. The experiment conditions were shown in Table 2.

Microstructure of the scale was observed by scanning electron microscope (SEM) and element distributions of copper and nickel in scale layer and steel substrate respectively were analyzed by energy dispersive X-ray spectra (EDS).

Table 1 Chemical composition of the continuous cast slabs used in experiment (mass percent, %)

Type of slab	C	Si	Mn	P	S	Ni	Cr	Al	Cu
Slab 1	0.180	0.260	0.88	0.014	0.005	0.040	1.120	0.0150	0.10
Slab 2	0.080	0.277	0.89	0.094	0.002	0.422	0.425	0.0635	0.52

Table 2 Experimental conditions for scale formation and Cu enrichment

Type of slab	Initial slab surface temperature, ℃	Continuously cooling condition
Slab 1	1100/1300	(1100/1300)℃→800℃ in H_2O vapor → room temperature in ambient air
Slab 2	1100/1300	(1100/1300)℃→800℃ in H_2O vapor → room temperature in ambient air

Results and Discussion

Microstructure of the Formed Scale on Continuously Cast Slab

Figure 1 shows the scale microstructure of Slab 1 after continuous cooling from different initial slab surface temperatures (1100℃ or 1300℃) to 800℃ in H_2O vapor atmosphere, and then consecutively cooling in air to room temperature. The scale microstructures of Slab 2 under the same cooling condition are shown in Fig. 2. It is noted that all the oxide scales show a relatively adherent to the steel substrate irrespective of the different initial surface temperatures and slab types, but a loosen and porous microstructure formed under the condition of higher initial slab surface temperature (1300℃), as seen both in Figs 1(b) and 2(b). This type of loosen and porous scale is considered to be caused by the volume expansion stress due to the phase transformation with decomposition from wustite to magnetite during cooling process. The stress increases at higher initial slab surface temperature because the wustite phase becomes more dominant in the scale layer. The tight adhesion and microstructure (dense/loosen) of scale is expected to have a significant influence on copper enrichment behaviour [4, 8].

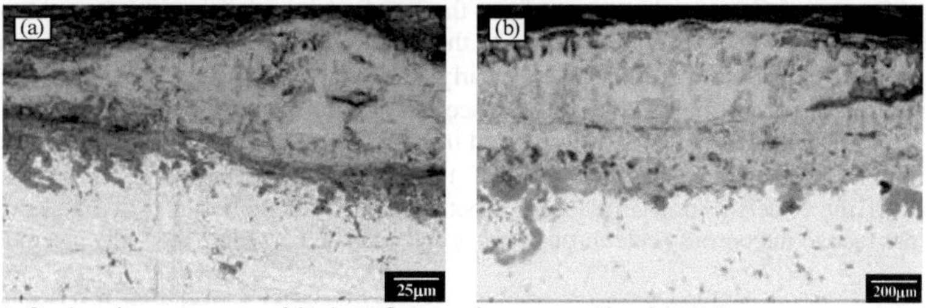

Fig. 1. Scale microstructure of Slab 1 after continuous cooling in H_2O vapor + consecutively in air for different initial surface temperatures. (a) 1100℃. (b) 1300℃.

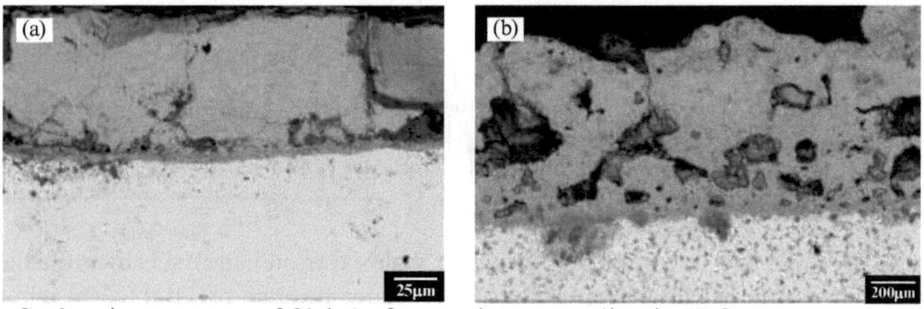

Fig. 2. Scale microstructure of Slab 2 after continuous cooling in H_2O vapor + consecutively in air for different initial surface temperatures. (a) 1100℃. (b) 1300℃.

Copper Enrichment Induced by Scale Formation

Figure 3 shows the line-scan profiles of copper element along scale layer and steel substrate of Slab 1 sample after continuous cooling in H_2O vapor atmosphere and consecutively in ambient air. Fig. 3(a) is the sample of Slab 1 with initial surface temperature of 1100℃ and Fig. 3(b) is with the initial surface temperature of 1300℃. No copper enrichment is observed either inside scale layer or at scale-substrate interface regardless of the surface temperature, indicating that for the Slab 1 samples with low copper content, as the scaling proceeds, the copper accumulation rate is not high enough to form notable copper enrichment at the scale/steel interface. Accordingly, the dense and loosen scales in Fig. 1 have not any influence on the copper enrichment. Thus, the recommendation to maintain a copper content less than 0.1 percent is considered to be important to reduce copper enrichment and hot shortness occurrence [11].

On the other hand, line-scan profiles of copper and nickel elements in Slab 2 samples are shown in Fig. 4. For the sample of Slab 2 with lower initial surface temperature (1100℃), the copper enrichment is found at scale-substrate interface, as shown in Fig. 4(a). Meanwhile, no notable nickel enrichment is observed in Slab 2 at 1100℃, as shown in Fig. 4(b). It suggests that for the slab with high contents of copper and nickel but low surface temperature, the oxidation rate is lowered because of the lower temperature and the presence of Ni, and the scaling rate is not rapid enough to expel copper and nickel simultaneously due to the higher oxidation potential of Ni. Thus, diffusion of Cu-rich phase away from the interface into the matrix is possible since the formed scale (Fig. 2(a)) is tightly attached to the substrate. While for the sample of Slab 2 with higher surface temperature (1300℃), it is clearly noted that Cu-enrichment inside the scale layer but a little away from the steel-scale interface, and simultaneously Ni-enrichment appears in front of the steel-scale interface, as illustrated in both Figs. 4(c) and (d). A porous scale formed at high initial slab surface temperature (1300℃) and positive effect resulting from Ni enrichment in front of the scale/substrate interface are both regarded to promote Cu-rich-phase migration from interface to the porous scale through pores and occlude Cu within the scale layer [1, 8].

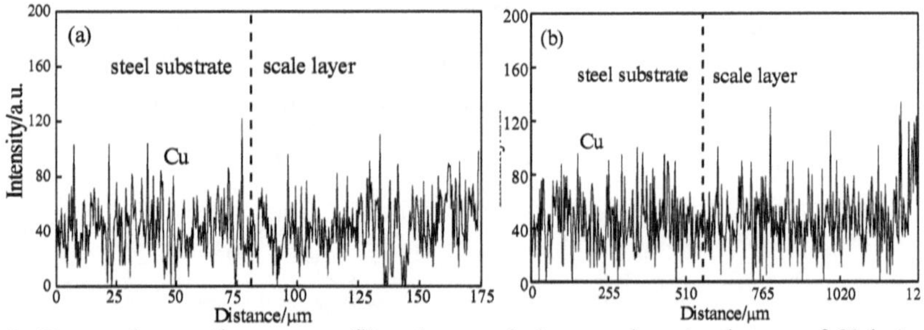

Fig. 3. Copper element line scan profiles along scale layer and steel substrate of Slab 1 with different slab surface temperatures. (a) initial surface temperature of 1100℃. (b) initial surface temperature of 1300℃.

Figure 5 shows the SEM image and EDS analysis result of Slab 2 sample with the initial surface temperature of 1300℃. It is clearly visible that copper-rich phase particles are occluded into the inner layer of oxide scale, which is consistent with the line-scan result of copper element in Fig. 4(a). Compared with that of Slab 2 sample at 1100℃, the different copper-enrichment behaviour in Slab 2 with initial surface temperature of 1300℃ reveals that with respect of the influence of scale microstructures (dense/porous scale formed at low/high initial slab surface temperature), the critical contents of copper and nickel in slab and initial surface temperature are also required simultaneously for the occlusion mechanism of copper into scale layer to operate. During continuously casting process, under the oxidizing atmosphere condition of water vapour in secondary cooling zone, scale formation promotes Cu accumulation at scale-substrate interface as oxidation proceeds, whereas a porous scale at higher initial slab surface temperature and high nickel content work together to occlude the copper-rich phase within the scale layer.

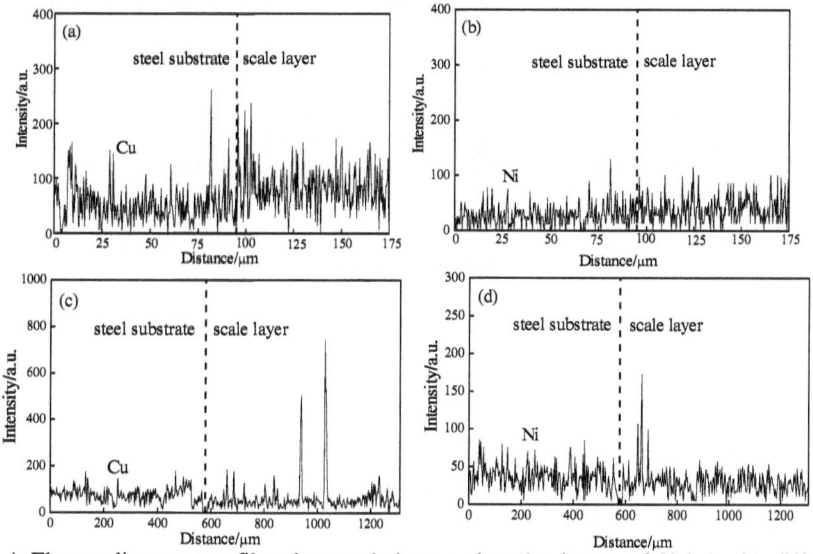

Fig. 4. Element line scan profiles along scale layer and steel substrate of Slab 2 with different surface temperatures. (a) Cu element at 1100℃, (b) Ni element at 1100℃, (c) Cu element at 1300℃, (d) Ni element at 1300℃.

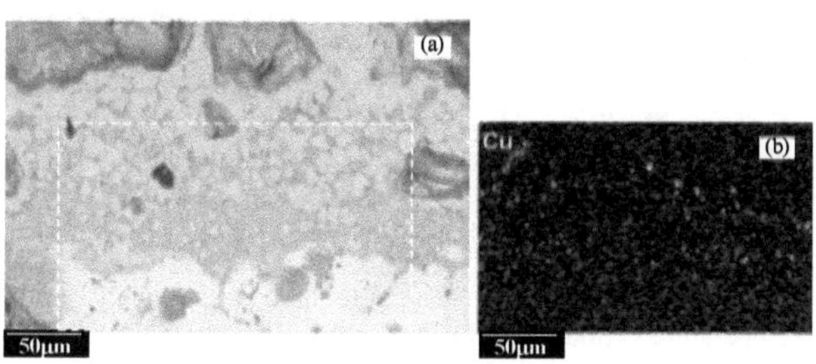

Fig. 5. Copper-rich phase occluded in the scale layer. (a) SEM image. (b) EDS map.

Conclusions

(1) A scale layer tightly adherent to the steel substrate is formed on continuously cast slab surface after continuous cooling from different initial slab surface temperatures in H_2O vapor atmosphere and consecutively in air. Compared with the dense scale formed at lower initial slab surface temperature, the scale microstructure is loose and porous at higher temperature. The different scale microstructures have significant effects on copper enrichment behaviour.

(2) For Slab 1 with lower copper content, no visible copper enrichment is found at scale-substrate interface regardless of the initial slab surface temperature. A copper content level less than 0.1 percent is recommended to be proper to reduce copper enrichment induced by scale formation.

(3) For Slab 2 with high contents of copper and nickel, under the condition of lower initial slab surface temperature, copper-rich phase accumulates and stays at scale-substrate interface with a formed dense scale tightly attached to the substrate. While at higher initial slab surface temperature, Cu-rich phase in the porous scale layer is noted with simultaneous Ni-enrichment appearing in front of the steel-scale interface. Both a porous scale formed at higher initial slab surface temperature and high nickel content act together to occlude the copper-rich phase within the scale layer.

Acknowledgments

This research work was supported by the National Natural Science Foundation of China (Grant No. 51174052, 51374062, 51574066, 51174049, 51374057)

.

References

[1] M. M. Wolf, “Scale Formation and Descaling in Continuous Casting and Hot Rolling: IV”, *I & SM*, 27 (4) (2000), 58-60.

[2] N. Imai, N. Komatsubara, K. Kunishige, “Effect of Cu, Sn and Ni on Hot Workability of Hot-rolled Mild Steel”, *ISIJ International*, 37 (3) (1997), 217-223.

[3] H. Matsuoka, K. Osawa, M. Ono, M. Ohmura, “Influence of Cu and Sn on Hot Ductility of Steels with Various C Content”, *ISIJ International*, 37 (3) (1997), 255-262.

[4] Y. Kondo, “Behaviour of Copper during High Temperature Oxidation of Steel Containing Copper”, *ISIJ International*, 44 (9) (2004), 1576-1580.

[5] C. Nagasaki, H. Uchino, K. Shibata, K. Asakura, M. Hatano, “Surface Hot Shortness due to Copper in IF Steel and Effect of Boron and Phosphorus ”, *Tetsu-to-Hagané*, 89 (3) (2003), 322-328.

[6] K. Shibata, S. Seo, M, Kaga, H. Uchino, A. Sasanuma, K. Asakura, C. Nagasaki, “Suppression of Surface Hot Shortness due to Cu in Recycled Steels”, *Mater. Trans.*, 43 (2002), 292-300.

[7] R. Y. Chen, W. Y. D. Yuen, “Copper Enrichment Behavior of Copper-containing Steels in Simulated Thin-slab Casting Processes”, *ISIJ International*, 45 (6) (2005), 807-816.

[8] S. W. Kim, H. G. Lee, “Effect of Oxide Scale Formation on the Behaviour of Cu in Steel during High Temperature Oxidation in O_2-N_2 and H_2O-N_2 Atmosphere”, *Steel Res Int*, 80 (2) (2009), 121-129.

[9] T. Kajitani, M. Wakoh, N. Tokumitsu, S. Ogibayashi, S. Mizoguchi, “Influence of Heating Temperature and Strain on Surface Crack in Carbon Steel Induced by Residual Copper”, *Tetsu-to-Hagané*, 81 (3) (1995) 29-34.

[10] N. Wang, J. H. Dong, W. J. Huang, B. Li, M. Chen, “Growth Rate, Microstructure and Phase Composition of Oxide Scales for Three Typical Steels in Simulated Continuous Casting Process” *J Iron Steel Res Int*, 21 (12) (2014), 1065-1072.

[11] M. M. Wolf, “Scale Formation and Descaling in Continuous Casting and Hot Rolling: Part I”, *I & SM*, 27(1) (2000), 22.

EPD Congress 2016
Edited by: Antoine Allanore, Laura Bartlett, Cong Wang, Lifeng Zhang, and Jonghyun Lee
TMS (The Minerals, Metals & Materials Society), 2016

PHYSICAL SIMULATION OF CRITICAL BLOWING RATE OF SLAG ENTRAPMENT OF 80 TONS LADLE

Rui Wang[1], Yanping Bao[1], Yihong Li[2], Aichun Zhao[2], Yafeng Ji[2], Xiao Hu[2], Qingxue Huang[2], Jiansheng Liu[2]

[1]State Key Laboratory of Advanced Metallurgy
University of Science and Technology Beijing
Beijing, 100083, China
[2] School of Materials Science and Engineering
Taiyuan University of Science and Technology
Taiyuan, 030024, China

eywords: Critical Blowing Rate; Slag Entrapment; Ladle Refining; Nozzle Arrangement

Abstract

The slag entrapment under different conditions of 80t blowing argon ladle furnace was investigated by physical simulation. The water was used to simulate liquid steel and liquid paraffin was for slag. The processing of slag entrapment under different blowing structures was analyzed and the critical velocity and critical droplets diameter of describing it was obtained. Based on the experiments, the relationship between the interface flow velocity and the critical blowing rate（CBR） was deduced. In the real process, it is suggested that the bottom blowing rate is from 40 L/min to 180L/min when the interface tension is 0.12~1.2 N/m during the soft argon blowing.

Introduction

Blowing argon of ladle is the effective way to improve the quality of steel, which has been applied widely in steel making industry. It is pivotal to control the blowing rate of argon, once it is above the critical value of slag entrapment, it will entrap slag into steel rather than cleaning the steel. Therefore, only the proper blowing rate can improve the reaction between slag and steel, uniform the composition and temperature, remove the inclusions in steel[1-3].

So far, there are many studies on the behavior of slag entrapment and factors of causing the steel exposed to the air. In order to avoid it and reduce the exposed area, the critical blowing rate(CBR) of slag entrapment is an important operating parameter[4-7]. However, there are different views about the results of CBR and few studies about the rule of critical slag entrapment[8-9]. In this study, to provide reference of blowing argon to the actual production, the CBR was obtained by physical simulation.

Experimental Method

The 80t ladle was taken as prototype for the research and the model of the ladle was constructed with a scale factor of 1:4. The N_2 was used to simulate argon and the water was simulated steel. Table 1 shows the main parameters of model and prototype of lade.

For the ladle refining system of blowing argon, the force causing the steel flowing is the buoyancy of bubbles, so under the condition of geometric similarity, the dynamic similarity can be obtained by equal modified Froude number of model and prototype. Due to $Fr_m' = Fr_p'$, the relationship of gas flowrate between the model and prototype was derived:

$$Q_m = 3.038\times10^{-2} Q_p \tag{1}$$

Where, Q_m and Q_p are respectively the gas flow rate of prototype and model.

The flow of steel-slag interface is mainly influenced by the interface tension, so the Weber number must be taken into consideration [10] .

According to the equal *We*, the relation of density between slag and the simulated slag is derived:

$$\rho_{ms} = \rho_w - \frac{\rho_{Ar}^2 \rho_w^4 \sigma_{s\text{-}s}^2}{\rho_{N_2}^2 \rho_s^4 \sigma_{w\text{-}ms}^2} \lambda^2 (\rho_s - \rho_{slag}) \tag{2}$$

Where, ρ_w, ρ_s, ρ_{ms} and ρ_{slag} are respectively the density of water, steel, simulated slag and slag, kg/m^3.

It is calculated that the density of simulated slag (ρ_{ms}) is 842 kg/m^3, which is very close to the density of liquid paraffin (850 kg/m^3). So the liquid paraffin dyed red by Sudan Red is chosen to simulate the slag so as to be easier observed.

In the experiment, the liquid paraffin is put into ladle slowly and keeps afloat above the water, of which the thickness meets the experimental requirement. To find out the conditions of slag entrapment, the different situations of steel-slag interface are recorded by camera under various blowing rate of nitrogen. The nitrogen injects into the bath through the air brick and the liquid in bath is uplifted with the rising bubbles, which forms the ridgy area above the steel-slag interface. The ridgy liquid flows downward by gravity causing the fluctuation of water-paraffin interface which becomes larger with the increasing of flowing rate. When the flow rate reaches a value, the drops will be formed at the interface. At this time, the value of gas flowrate is called the CBR of slag entrapment.

Table I Main parameters of prototype and model

Parameter	Upper diameter (mm)	Bottom diameter (mm)	Bath height (mm)	Gas density (m^3/h)	Liquid density (m^3/h)	Surface tension (N/m)
Prototype	2650	2300	2600	1.79	7000	1.22
Model	662.5	575	650	1.25	1000	0.042

Results and Discussions

The ualitative description of slag entrapment

Fig.1 shows the variation of simulated slag entrapment with the incensement of blowing rate when the gas injects through single air brick. The mild fluctuation occurs at paraffin-water interface when the blowing rate is small, and the gas discharges through the liquid paraffin (as shown in fig.1a); when the blowing rate become larger, as shown in fig.1b, there is emerging a ridgy area at the paraffin-water interface and there is no paraffin drops breakup; while the blowing rate is increasing as shown in fig.1c and fig.1d, the distinct fluctuation appears at interface and the liquid paraffin layer was blown a round area where the water exposed to air. There are few numbers of scattered paraffin drops is involved in water but quickly ascents back to paraffin layer; as the blowing rate is increasing continuously, as shown in fig.1e and fig.1f, there are more paraffin drops scattering deeply in the water and they are difficult to rise back to the interface. The above phenomenon is consist with the description of slag entrapment in Chen's study [11].

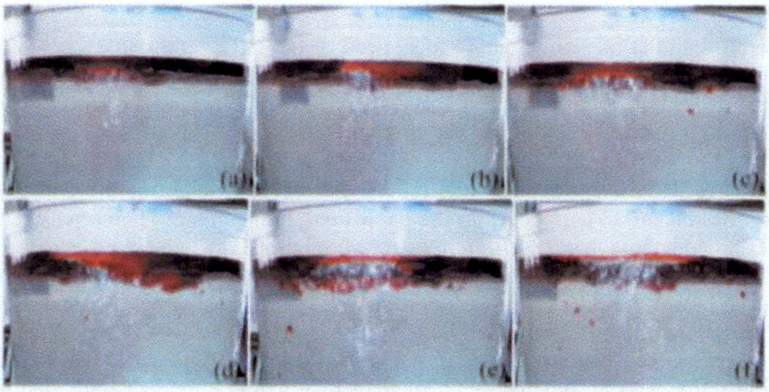

Fig. 1 Pictures of liquid paraffin-water interface at different gas flowrates with a single air brick

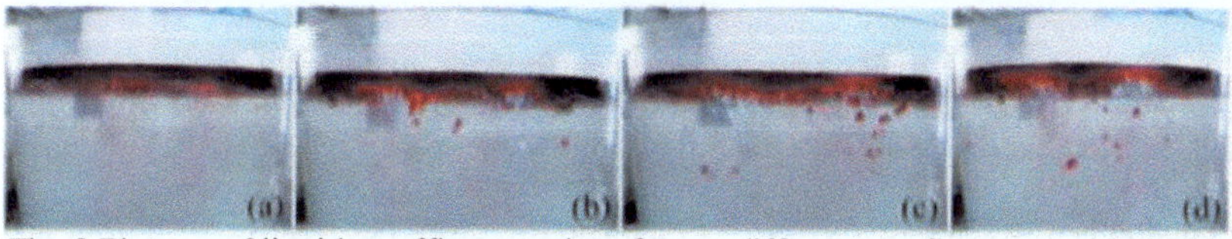

Fig. 2 Pictures of liquid paraffin-water interface at different gas flowrates with double air bricks

In fig.2, it gives the variation of simulated slag entrapment under different bottom blowing rate through two air bricks. With the increase of blowing rate, the phenomenon is similar with that blowing with single air brick. The blowing rate in fig.2a, fig.2b, fig.2c and fig.2d is respectively matching that in fig.1c, fig.1d, fig.1e and fig.1f. Compared with fig.1, under the same bottom blowing rate, the fluctuation in fig.2 is smaller than that in fig.1; the diameter of scattered drops in fig.2 is larger than that in fig.1 but the depth that the drops reach in fig.2 is smaller. That is the scattered drops with blowing through two bricks are liable to come back to the paraffin-water interface. And the blowing rate causing the paraffin drops through two air bricks is greater than that through single air brick.

Slag drops entrapment of bottom blowing

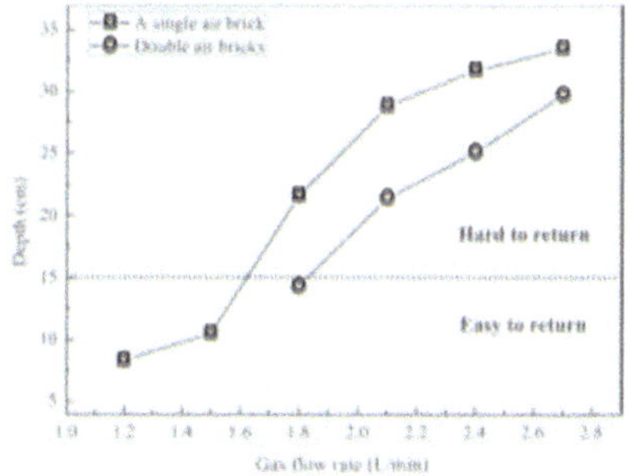

Fig. 3 Effect of gas flow rate on the depth of droplets into the water

Fig.3 shows the depth of slag drops in bath influenced by bottom blowing rate of argon. When gas flow rate is below 1.8L/min, the depth of slag drops entrapped into bath varies slightly with rising of gas rate. It is becuase the inertial force of drops is smaller than the buoyancy due to low speed of it. And Zheng's study[12] shows that the liquid drops can be float up to interface by its own buoyancy when the immersion depth is below 15cm.And it quickly increases to 22cm at the gas flow rate of 1.8L/min,with ascent of bottom blowing rate the speed of liquid is become larger, the circulation velocity of liquid aroud gas column enlarges and the drops velocity to bottom is also increasing. The inertial force of drops is larger than the buoyancy, so the drops is hard to come back to the paraffin-water interface. Therefore, the gas flow rate should be less than 1.8L/min in oder to littel drops involved in water.

It is the similar results that the immersion depth of slag drops influenced by bottom blowing rate both through two air bricks and single one. But under the same blowing rate, the immersion depth of slag drops is smaller with blowing through two air bricks, which is consistent with the situation in fig.1 and fig.2. It is because that the injection area of two air bricks is larger than that of single air brick, the injection speed through two air bricks is lower, so the circulation speed of liquid is lower and the slag drops of inertia force under two air bricks is smaller, causing the smaller immersion depth.

There are three forces exerted on a paraffin drop in the process of the paraffin drop fall from the parafiin layer[11], respectively inertia forece F_i coming from fluid flow, interfacial tension F_s and the buoyancy F_b due to density difference. Assumed that the angle between the entrapment direction of paraffin drops and vertical direction is α, when $F_i \geq F_b/\cos\alpha + F_s$ the paraffin drops fall from paraffin layer and surbmerges in the bath; when $F_i = F_b/\cos\alpha + F_s$, the paraffin drops is in the critical state of slag entrapment. Therefore, when α=0° , it is easiest to appear slag entrapment because of the minimum intertia force needed under this condition; when α=90° the directions of inertia force and interfical tension exerted on drops is horizontal and opposite, so the paraffin drop can float up to the interface by buoyancy rather than entrapment.

According to the condition of slag entrapment, literature [11] gives ciritical velocity of slag entrapment and critical diameter of slag drops respectively shown in Formula (3) and (4):

$$u_{cr} = 2\sqrt{\frac{\sqrt{6\sigma g(\rho_s - \rho_{slag})\tan\alpha}}{\rho_{slag}}} \tag{3}$$

$$d_{cr} = \sqrt{\frac{3\sigma}{g(\rho_s - \rho_{slag})\tan\alpha}} \tag{4}$$

It is assumed that the paraffin drop generates at the intersection of gas column and paraffin-water interface and the direction of slag entrapment is along the tangent of interface, so the α can be obtained by width of gas column under different blowing rate. Table II calculates the critical velocity and critical diameter of slag entrapment according to formula (3) and (4).

Table II The value of critical velocity and critical diameter of droplets

The width of air column (m)	α (°)	Critical velocity (m/s)	Critical diameter of droplets (mm)
A single air brick	81.2	0.482	2.48
	80.5	0.473	2.57
	78.0	0.446	2.91
	77.4	0.441	2.98
	76.0	0.428	3.15
	74.8	0.419	3.28

Double air bricks	78.7	0.455	2.79
	78.3	0.449	2.86
	78.1	0.446	2.89
	77.6	0.441	2.96
Actual measurement	47	0.307	6.10
	30	0.264	8.30

From table 2, in this study, it is difficult to appear slag entrapment when the α is near 90°; but the probability of it is increasing with decreasing of α and the low critical velocity cause to large critical diameter of paraffin drops. So as to avoid slag entrapment, the interface rate must be smaller than that in Table 2. However, in practical production, the interface rate only can be controlled by bottom blowing rate. So it is important to find out the relationship between argon blowing rate and interface rate, which can give full play to clean steel by blowing argon.

The relationship between critical velocity and critical blowing rate

The kinetic energy of paraffin drops separating from slag is from the kinetic of blowing gas by water (steel) transmitting. It is supposed that u_w is water velocity and u_s is velocity of paraffin layer. Due to little diameter of paraffin drops formed at interface, it is considered that the velocity of interface is equal to the velocity of paraffin layer. According to equivalent shear force of paraffin and water, Oersted [13] deduced the formula of dimensionless number $U=u_i/u_w$:

$$U = 0.1367(\frac{\rho_w}{\rho_{ms}})^{2/3} \cdot (\frac{u_w l}{\nu_{ms}})^{1/3} \cdot (\frac{u_w l}{\nu_w})^{-2/15} \cdot [(1-U)(0.1108-0.0693U)]^{2/15} \tag{5}$$

Where, u_w and u_i is respectively the interface velocity of water and interface, m/s; ν_{ms} and ν_w is the kinematic viscosity of paraffin and water, respectively 1.0×10^{-6} m^2/s and 34.5×10^{-6} m^2/s; l is the length of paraffin-slag interface exerting shear force on, $l=H_{ms}/\cos\alpha$, m; H_{ms} is the thickness of simulated slag, namely thickness of paraffin layer, in this study, the thickness is 0.03m.

Below the interface, the velocity of water u_w is equal to the liquid velocity u_b in gas column, above the interface it is accelerating by the gravity [15].

$$u_w = \sqrt{u_b^2 + gH_{ms}} \tag{6}$$

$$u_b = 1.17(\frac{Q_g H_w}{A_b})^{0.346} \tag{7}$$

Where, u_b is water velocity in the gas column, m/s; Q_g is bottom blowing rate, m^3/h; H_w is the depth of bath, m; A_b is the gas-liquid area of cross-section, m^2.

The interface rate of paraffin drops is calculated according to formula (5)~(7). In the calculation, the fluctuation of interface, namely the thickness of paraffin layer, affected by fluctuation is taken into consideration. The thickness paraffin layer is, the easier to form paraffin drops at the thickest paraffin layer. In this study, the thickness of simulated slag layer is 0.03m and its maximum thickness is 0.06m during the bottom blowing. Compared the critical interface rate obtained by theoretical calculation in table 2 with that deduced by experimental flowrate, it shows that they matches each other in fig.4. It also concluded that the critical blowing rate instead of critical interface rate can be used to estimate the slag entrapment, and there is a relationship between them.

Combined with conditions of this study, the dimensionless number (U=0.55) is calculated from formula (5)~(7), accordingly, $u_i=0.55u_w$.

$$Q_g^{0.692} = \left(2.41u_i^2 - 0.73gH_{ms}\right) \cdot \left(A_b / H_w\right)^{0.692} \tag{8}$$

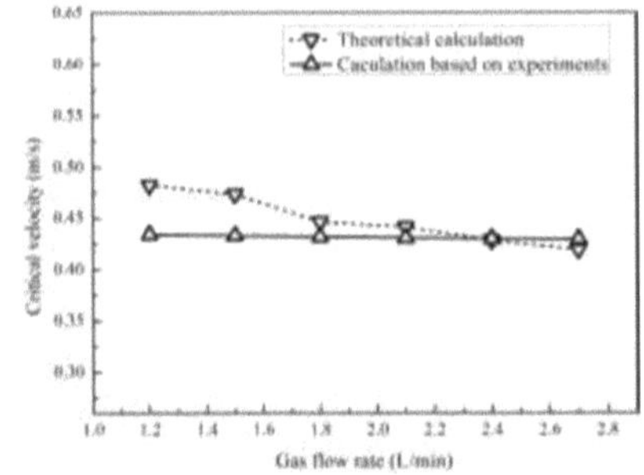

Fig. 4 Morphology of typical inclusions at the beginning of LF

Under the maximum interface rate of this study, the critical bottom blowing rate are respectively 1.49 L/min and 1.70 L/min through single air brick and two air bricks. A more narrow range of critical gas blowing rate is obtained by the above conclusion.

Critical flowing rate in practical

For the slag-steel system of ladle, the Weber number is the definite number describing the state of the system [16]. According to similar principle, the interface rate of steel is derived:

$$We_{\text{liquid paraffin/water}} = We_{\text{slag/steel}} = \frac{U_s^2 \cdot \rho_s}{[\sigma_{s\text{-}s} \cdot g(\rho_s - \rho_{\text{slag}})]^{1/2}} = 22.6 \tag{9}$$

$$U_s = \sqrt{\frac{22.6 \cdot [\sigma_{s\text{-}s} \cdot g(\rho_s - \rho_{\text{slag}})]^{1/2}}{\rho_s}} \tag{10}$$

In literature [17], the slag-steel interface tension of steady state is 1.2N/m and it varies from 0.12N/m to 1.2 N/m when in interface reaction and in unstable state. The flow velocity of steel at the interface is 0.458~0.814m/s by formula (10). Take the results into formula (8), the blowing rate in practical production is 40~180L/min when the interface is unsteady.

Due to the unfinished reconstructing of ladle, the industry test is to be carried out in the coming.

Conclusion

(1) The slag entrapment in steel refining is influenced by the bottom argon blowing rate definitively; when the blowing rate increased to 1.8L/min the depth of paraffin falling into the bath largely increases. To avoid slag entrapment, the blowing rate should be controlled below 1.8L/min. It is more suggestion that the bottom blowing rate should be below respectively 1.49 L/min and 1.70L/min through single air brick and two air bricks.

(2) The critical blowing rate instead of critical interface rate can be used to estimate the slag entrapment when the thickness of paraffin changes due to fluctuation of interface. The relationship between them is shown as: $Q_g^{0.692}=(2.41u_i^2 -0.73gH_{ms})(A_b / H_w)^{0.692}$.

(3) The critical blowing rate in practical is influenced by temperature and instability of interface reaction and it is suggestion that the blowing rate of argon is controlled from 40 L/min to 180 L/min during soft blowing.

REFERENCES

1. Xiao Z Q, Liu C L, Hu L X, et al. "Behavior and Source of Ladle Inclusions in Steel Treated by Power Injection", *Iron and Steel*, 1988, 23(2):23-29.
2. Gupta D, Lahiri A K. "A water model study of the flow asymmetry inside a continuous casting mold", *Metall. Mater. Trans. B*, 1996, 27(5): 757-762.
3. Zheng S G, Zhu M Y. "Water model study on removing inclusions in a ladle with argon injected through nozzle and porous plug", *Acta Metallurgica Sinica*, 2006, 42(11): 1143-1148.
4. Jonsson L, Joensson P. "Modelling of fluid flow conditions around the slag/metal interface in a gas-stirred ladle". *ISIJ Int.*, 1996, 36(9): 1127-1134.
5. Han J W, Heo S H. "Transient fluid flow phenomena in a gas stirred liquid bath with top oil layer-approach numerical simulation and water mold experiment". *ISIJ Int.*, 2001, 41(10): 1165-1172.
6. Krishnapisharody K, Irons G A. "Modeling of slag eye formation over a metal bath due to gas bubbling". Metall. Mater. Trans. B, 2006, 37(5): 763-772.
7. Li B K, Gu M Y, Qi F S, et al. "Modeling of three-phase(gas/molten steel/slag) flows and slag layer behavior in an argon ags stirred ladle". *Acta Metallurgica Sinica*, 2008, 44(10): 1198−1202.
8. Mamabu I, Yutaka S, Ryusuke O, et a1. "Evaluation of critical gas flow rate for the entrapment of slag using a water model". *ISIJ Int.*, 1994, 34(2): 164-170.
9. Kim S H , Fruehan R J , Guthrie R I L. "Physical model studies of slag/metal reactions in gas sti rred ladles-determination of critical gas flow rate". *Iron & Steel Maker*, 1993 , 20(11): 71.
10. Zhu M Y, Xiao Z Q. "Mathmatical and physical simulation in steel refining". Beijing, Metallurgical Industry Press, 1998.
11. Cheng G G, Zhang J, Yi X J. "Study on mechansim of slag entrapment in bottom Ar-blowing for ladle". *Steelmaking*, 1993, (6): 23-25.
12. Zheng W, Tu H, Li G Q, et al." Modeling of slag entrapment and molten steel expsed to atmosphere in refining of 250 t ladle weith bottim-blown argon". *The Chinese Journal of Process Engineering*, 2014, 14(3): 361-367.
13. Franz Oeters, Metallurgy of Steelmaking ,Berlin, Verlag Stahleisen mbH, 1989.
14. Masamichi S, Kazumi M. "Fluid flow and mixing characteristics in a gas-sti rred molten metal bath". *Transact ions ISIJ*, 1983, 23: 169-175 .
15. Gan L, He P. "Study on the characteristics of critical gas flowrate for the entrapment of slag in gas-stirred ladles," *Steelmaking*, 2009, 25 (1): 17-21.
16. Xiao Z Q, Peng Y C. "Mathematical modelling of entrapment phenomena at slag/metal interface in gas stirred ladle," *IRON AND STEEL*, 1989, 24 (10): 41-46.
17. Szekely J, *Fluid flow phenomena in metals processing* (Academic Press, 1979).
18. Qu Y, *Theroy of Steelmaking* (Beijing, BJ: Metallurgical Industry Press, 1980).

EPD Congress 2016
Edited by: Antoine Allanore, Laura Bartlett, Cong Wang, Lifeng Zhang, and Jonghyun Lee
TMS (The Minerals, Metals & Materials Society), 2016

LIQUID METAL MODELLING OF FLOW PHENOMENA IN THE CONTINUOUS CASTING PROCESS OF STEEL

K. Timmel, B. Willers, T. Wondrak, M. Röder, N. Shevchenko, G. Gerbeth and S. Eckert

Helmholtz-Zentrum Dresden-Rossendorf (HZDR), Dresden, Germany

Keywords: Continuous Casting, liquid metal model, electro-magnetic flow control, Argon injection, flow measurement

Abstract

The quality of the produced steel in the continuous casting process is significantly governed by the melt flow in the mold. However, direct flow measurements in liquid metals are still rather scarce. In order to investigate these flow phenomena, three experimental facilities operating with low melting liquid metals were installed at Helmholtz-Zentrum Dresden-Rossendorf (HZDR). The melt flow in the models is measured by the Ultrasonic Doppler Velocimetry (UDV) or the Contactless Inductive Flow Tomography (CIFT), multi-phase flows can be visualized by X-ray imaging. The obtained measurement results are primarily used for validation of numerical models.

In this paper we will investigate the fluid flow in the mold and the behavior of the surface of the liquid metal using flow measurements by UDV and surface profile measurement by a laser scanner, respectively. Strong fluctuations and deviations of the free surface were observed in case of a static magnetic field.

Introduction

The flow of the liquid steel in the tundish, the submerged entry nozzle (SEN) and the mold is one of the major parameters impairing the steel quality at the end of the process. A lot of effort is made to ensure a good quality by adjusting and controlling the liquid steel flow. The steel flow can be influenced by plant design, i.e. the geometry of the SEN, but this is fixed throughout the operation the continuous caster. More flexible tools for flow control are electromagnetic fields, which are already in industrial use since about 30 years. There are basically two types of magnetic fields in operation: static magnetic fields which are assumed to brake the flow, and alternating magnetic fields for stirring and pumping the liquid metal in a specific direction. A further issue is the injection of argon into the SEN for removal of inclusions and the prevention of nozzle clogging. Despite their wide application in industrial plants, the actual action of the electromagnetic fields as well as the behavior and distribution of the argon bubbles is still not satisfyingly clarified and under discussion.

Several plant trials have been carried out [1-3]. Though, only coarse and insufficient measurement data can be retrieved from real steel casters. Because of the high temperature of the liquid steel and the harsh environment at the casting machine, there are almost no measurement techniques available. Some rough information might be retrieved from observations of the free surface by dipping paddles or nail-boards into the melt, but flow measurements from deep inside the liquid steel are still missing and there is no solution foreseeable for the next years.

During the last decades water experiments are used for modelling the flow in the SEN and the mold respective investigations of the flow field were done by optical methods, e.g. [4-7]. But some material properties of water differ dramatically from those of liquid metals, like the electrical conductivity or the surface tension. This difference leads to a serious discrepancy in the dimensionless numbers, such as Hartmann number or magnetic interaction parameter, which cannot be compensated by scaling other parameters. In the special cases of a two-phase flow and in particular in the case of applied electromagnetic fields, water experiments are no longer appropriate.

Such important phenomena can be only modelled correctly by liquid metal experiments. Owing to the limited availability of measurement techniques for liquid metals, only few experimental results are reported in the literature using setups operating with low melting liquid metals or alloys. First pioneering works used mercury as model liquid [8-10]. In the last decades large progress was made in the development of appropriate measurement techniques for liquid metals at moderate temperatures. So, a more detailed investigation of the continuous casting process using such liquid metal experiments are now possible. This paper will give an overview about three experimental facilities build up at the HZDR within the LIMMCAST program. Exemplary results and recent developments will be presented.

Such 'cold' liquid metal experiments on continuous casting become important because they provide a broad experimental data base for validation of numerical models, which can then be adapted to the huge variety of continuous casting geometries and operating conditions.

Experimental facilities

Three experimental facilities are available at the Helmholtz-Zentrum Dresden-Rossendorf for modeling the essential thermal hydraulic features of the continuous casting process using low melting liquid metals. The three setups have slightly different scopes of investigation, such as velocity measurement in single or two-phase flows, visualization of argon bubbles in the mold and SEN or the interaction of electromagnetic devices with the liquid metal flow. Various measurement techniques are applied to determine the flow field, e.g. the Ultrasound Doppler Velocimetry (UDV) with standard transducers or with ultrasonic wave guides, the Contactless Inductive Flow Tomography (CIFT), local potential probes or the X-ray radioscopy.

The main facility is the LIMMCAST (Figure 1(a)) setup operating with the $Sn_{60}Bi_{40}$ alloy at temperatures of 200 °C up to 350°C. The temperature-dependent material properties of the alloy can be found in literature [11]. The actual mold has a cross section of 400 x 100 mm^2. The facility was designed as flexible assembly with flange connections, which allows changing the important parts as the SEN or the mold in an adequate amount of time. A more detailed description can be found in [12].

The second facility is called Mini-LIMMCAST (Figure 1(b)), because it can be considered as small-scale copy of the big LIMMCAST model. Mini-LIMMCAST uses the eutectic GaInSn alloy as model liquid. The material properties are reported in the literature [13]. This alloy is liquid at room temperature and therefore no heating of the experimental setup is necessary. This simplifies the operation and the access of measurement techniques. This setup is also described in more details in [12]. Mini-LIMMCAST offers a continuous operation of diverse experiments with an active tundish level control.

The third facility is the X-LIMMCAST (Figure 1(c)). This setup focuses on the visualization of the two-phase flow in the SEN and the mold by X-ray imaging. The model liquid is again the eutectic GaInSn alloy. The thickness of the liquid metal in X-ray direction is limited due to the high attenuation of the X-rays by the liquid metal. The thickness of the mold and the SEN were therefore constricted to the maximum value of 15 mm. The SEN has a square cross section, to achieve a good resolution of the argon bubbles over the whole width and to avoid a saturation of X-ray imaging due to the continuously changing attenuation in a circular cross section. The setup of X-LIMMCAST is described in [17].

Experimental results

Previous studies at the Mini-LIMMCAST facility concerned the effect of an electromagnetic brake [18]. It was shown, that the static magnetic field gives rise to considerable changes of the mold flow which do not solely show the expected effect of an electromagnetic brake, namely a reduction of the highest velocity peaks. The application of a static magnetic field can lead to the formation of recirculation zones beside the jet and distinct oscillations of the jet position. An up and down bending of the jet was observed in the experiments, which was confirmed in later numerical simulations [19, 20]. The double roll flow structure in the mold is changed by the magnetic field to a multi roll flow structure. An additional roll right below the free surface may be created. A great single roll can develop below the SEN, which covers the whole mold width.

(a)

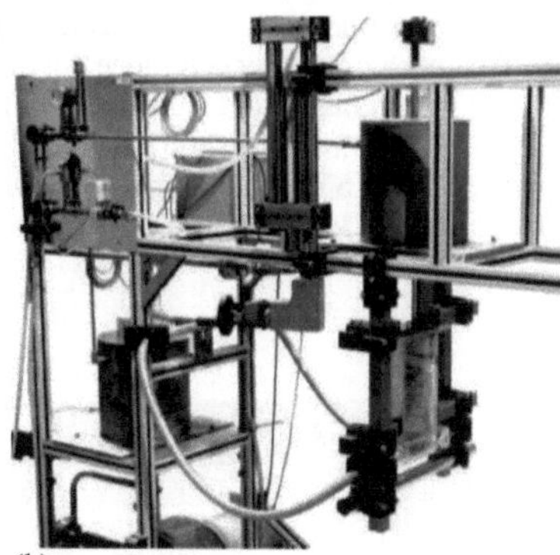

(b)

Figure 1: Fotos of the experimental facilities:
(a) LIMMCAST – large scale liquid metal experiment for continuous flow measurement in tundish, SEN and mold (250 °C),
(b) Mini-LIMMCAST – small scale acrylic glass model for experimental flow investigations at room temperature using GaInSn,
(c) X-LIMMCAST – small scale acrylic glass model for investigation of liquid metal-gas two-phase flow in the mold and SEN by X-ray imaging.

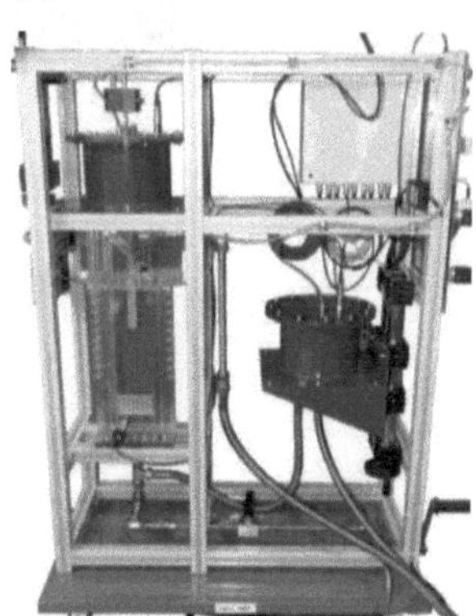

(c)

The rotating direction of the lower roll may change over time. These findings are in contrast to the idealized picture of a symmetric double roll structure in the mold. This temporal behavior of the jet and the roll structure is strongly dependent on the electrical boundary conditions.
The mold flow structure influences the behavior of the free melt surface. In continuous casting, highest attention is especially concentrated on the free surface in the mold because a violated interface between the liquid steel and the slag layer may impair the surface quality of the casting product. The relation between mold flow and surface behavior is therefor of peculiar interest. The application of any kind of magnetic field at the mold has to guarantee a stable and quiet surface.

The free liquid metal surface
Reference measurements at the Mini-LIMMCAST facility without magnetic field revealed the existence a standing wave at the free surface with a nodal point in the middle of the mold near the SEN. Wave peaks occurred at the narrow wall. The frequency of the standing wave is related to the mold width and corresponds to the natural frequency of a surface wave of twice the wavelength of the mold dimension. Surface waves were also observed in earlier mercury experiments [10]. The electromagnetic brake creates a strong upward flow along the narrow wall towards the free surface [20], which causes a strong bulging of the surface at this location. The surface behavior under reference conditions and under influence of an electromagnetic brake will be the next focus of detailed investigation at the Mini-LIMMCAST and the LIMMCAST setup. Figure 2 shows a snapshot from the free liquid metal surface at Mini-LIMMCAST (Fig. 2(a)) and LIMMCAST (Fig. 2(b)). The surface position will be measured by laser, ultrasonic methods, and immersed level sensors.

(a)
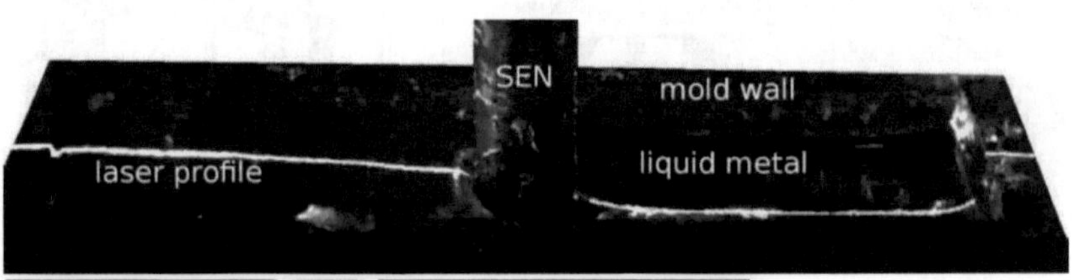

(b)
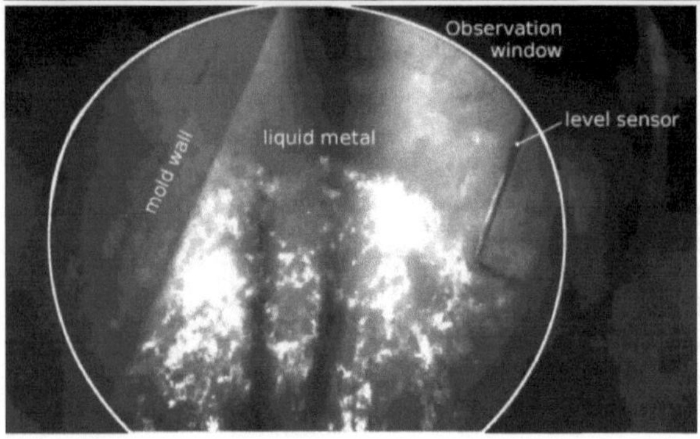

Figure 2: Surface visualization in the center plane by a line laser at Mini-LIMMCAST (a) and photograph of the free surface at LIMMCAST during an experiment without magnetic field (b).

A first example of the surface profile recorded by a line-laser distance measurement system under the conditions of a static magnetic field with electrically isolated mold walls is displayed in Figure 3(b). The line-laser system did not cover the whole mold half, but it gives a series of instant surface profiles with a scan rate of about 90 Hz. The whole mold width can be captured completely by two separate line-lasers on each mold half (as shown in Fig. 2(a)), equipped with a fast camera system and an independent, subsequent data analysis of the images. The bulging of the liquid metal surface is clearly visible. The bulge is located near the narrow wall, where a strong upward flow is triggered by the static magnetic field. This strong vertical flow is illustrated in Figure 3(a). The vertical velocity was measured by the Ultrasound Doppler Velocimetry at a width position of x = 55 mm (x = 70 mm corresponds to the position of the narrow wall). It can be clearly seen at this position near the narrow wall that the vertical flow above the jet (z > 0 mm) is forced upwards by the static magnetic field. In the reference case without magnetic field the flow shows a downward direction, indicating that the upward flow of the upper roll is obviously restricted closer to the narrow wall.

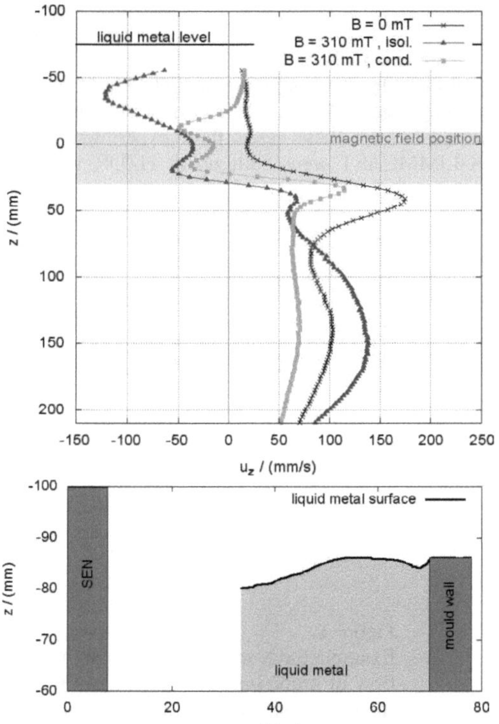

(a)

(b)

Figure 3:
Magnitude of vertical velocity u_z at x = 55 mm (a) and a snapshot of surface profile measurement by a line laser system in case of an applied magnetic field with isolating mold walls (b).

Two-phase flow in the SEN

Figure 4 gives an illustration of another important aspect of the continuous casting process: the two-phase flow in the submerged entry nozzle resulting from argon gas injection at the tip of the stopper rod. The snapshot of the two-phase flow was taken at the X-LIMMCAST setup by X-ray imaging. It shows clearly a complex flow regime with a bubbly flow and the creation of great void zones at the beginning of the SEN in the dead water zone of the inlet flow. Most of the released bubbles from the stopper rod tip are captured by the big void zones, although some of them have the chance to be dragged with the liquid metal flow through the narrow gap between the void zones, as shown in Fig. 4. New bubbles are sheared off from the end of the void zones and dragged downwards into the mold. Once created, the void zones are rather stable and remain for a long time, even when the argon is turned off.

The snapshot makes also clear, that it is quite difficult to distinguish single bubbles in the image. The flow in the SEN is very fast (in the order of 1 m/s) and bubbles are smeared during the exposure so that it is not possible to get a clear image of the bubble shape. On the other hand the smearing allow for an estimation of the bubble velocity in the SEN. New equipment has been installed to improve the image quality as well as the temporal resolution. First trials at other experimental setups were successfully done and the new equipment will be tested at X-LIMMCAST in the near future.

Summary

For the physical modeling of the continuous casting process, it is essential to use liquid metals for conducting the experiments when regarding special effects, e.g. in case of two-phase flows or magnetic fields. To provide an experimental tool for these cases, three experimental facilities LIMMCAST, Mini-LIMMCAST and X-LIMMCAST were built up at HZDR with slightly

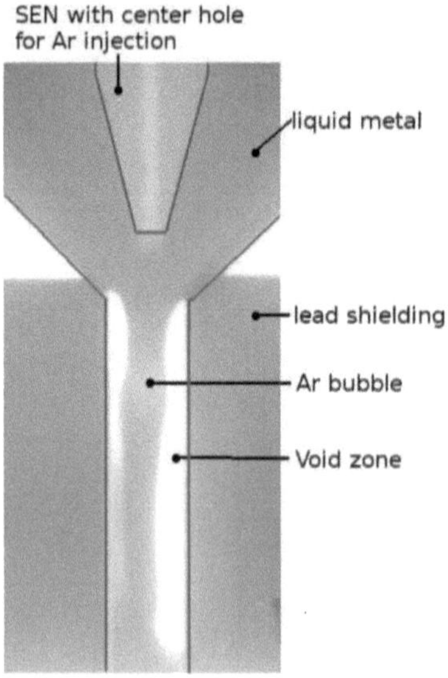

Figure 4:
Example for X-ray imaging of the two-phase flow by argon injection at the tip of the stopper rod in the upper part of the SEN. Great void zones are created at the inlet.

different scopes. This paper shortly presented all three experimental setups and exemplarily some measurement example. It could be demonstrated, that a static magnetic field may not obligatorily lead to a braking action in the liquid metal. At some points the flow may even be accelerated. The change in flow field due to the magnetic field has also consequences for the shape of the free liquid metal surface. X-ray imaging revealed a rather complex two-phase flow in the experiment. The measurement results of the liquid metal experiments represent a valuable data base for the validation of numerical models, e.g. [19-22].

Acknowledgment

The authors acknowledge the financial support from the German Helmholtz Association in the framework of the Helmholtz Alliance LIMTECH.

References

1. P. Gardin, J.-M. Galpin, M.-C. Regnier, J.-P. Radot, „Influence of electromagnetic brake on molten steel flow and inclusion behaviour in a continuous casting mold", *Magnetohydrodynamics*, 32 (1996), 189-195.
2. R. M. McDavid, B. G. Thomas, "Flow and thermal behavior of the top surface flux/powder layers in continuous casting molds", *Metall. Mater. Trans. B*, Springer Boston, 27 (1996), 672-685
3. K.H. Moon, H.K. Shin, B.J. Kim, J.Y. Chung, Y.S. Hwang, J.K Yoon, „Flow control of molten steel by electromagnetic brake in the continuous casting mold ", *ISIJ Int.*, 36 (1996), S201-203.
4. H. Bai, B. G. Thomas, "Turbulent flow of liquid steel and argon bubbles in slide-gate tundish nozzles: Part I. model development and validation", *Metallurgical and Materials Transactions B*, 32 (2001), 253-267
5. Q. Yuan, S. P. Vanka, B. G. Thomas, S. Sivaramakrishnan, "Computational and experimental study of turbulent flow in a 0.4-scale water model of a continuous steel caster", *Metall. Mater. Trans. B*, 35 (2004), 967-982
6. D. Gupta, A. Lahiri, "Water-modeling study of the surface disturbances in continuous slab caster", *Metall. Mater. Trans. B*, 25 (1994), 227-233
7. R. Kalter, M. Tummers, S. Kenjeres, B. Righolt, C. Kleijn, "Oscillations of the fluid flow and the free surface in a cavity with a submerged bifurcated nozzle", *International Journal of Heat and Fluid Flow*, 44 (2013), 365-374
8. K. Okazawa, T. Toh, J. Fukuda, T. Kawase, M. Toki, "Fluid flow in a continuous casting mold driven by linear induction motors", *ISIJ Int.*, 41 (2001), 851-858.
9. H. Harada, T. Toh, T. Ishii, K. Kaneko, E. Takeuchi," Effect of magnetic field conditions on the electromagnetic braking efficiency", *ISIJ Int.*, 41 (2001), 1236-1244.
10. J. Etay, Y. Delannoy, "Low frequency wave at the meniscus of a continuous caster generated by a dc magnetic field", *Magnetohydrodynamics*, 39 (2003), 445-451
11. Y. Plevachuk, V. Sklyarchuk, A. Yakymovych, B. Willers, S. Eckert, "Electronic properties and viscosity of liquid Pb-Sn alloys", *Journal of Alloys and Compounds*, 394 (2005), 63-68
12. K. Timmel, S. Eckert, G. Gerbeth, F. Stefani, T. Wondrak, "Experimental modeling of the continuous casting process of steel using low melting point metal alloys -- the LIMMCAST program", *ISIJ Int.*, 50 (2010), 1134-1141

13. Y. Plevachuk, V. Sklyarchuk, S. Eckert, G. Gerbeth, R. Novakovic, "Thermophysical properties of the liquid Ga-In-Sn eutectic alloy", *Journal of Chemical & Engineering* Data, 59 (2014), 757-763
14. T. Wondrak, V. Galindo, G. Gerbeth, T. Gundrum, F. Stefani, K. Timmel, "Contactless inductive flow tomography for a model of continuous steel casting", *Measurement Science and Technology*, 21 (2010), 045402
15. N. Terzija, W. Yin, G. Gerbeth, F. Stefani, K. Timmel, T. Wondrak, A. J. Peyton, "Use of electromagnetic induction tomography for monitoring liquid metal/gas flow regimes on a model of an industrial steel caster", *Measurement Science and Technology*, 22 (2011), 015501
16. T. Wondrak, S. Eckert, G. Gerbeth, K. Klotsche, F. Stefani, K. Timmel, A. J. Peyton, N. Terzija, W. Yin, "Combined electromagnetic tomography for determining two-phase flow characteristics in the submerged entry nozzle and in the mold of a continuous casting model", *Metall. Mater. Trans. B*, 42 (2011), 1201-1210
17. K. Timmel, N. Shevchenko, M. Röder, M. Anderhuber, P. Gardin, S. Eckert, G. Gerbeth, "Visualization of Liquid Metal Two-phase Flows in a Physical Model of the Continuous Casting Process of Steel" , *Metall. Mater. Trans. B*, 46 (2015), 700-710
18. K. Timmel, S. Eckert, G. Gerbeth, "Experimental investigation of the flow in a continuous-casting mold under the influence of a transverse, direct current magnetic field", *Metall. Mater. Trans. B*, 42 (2011), 68-80
19. R. Chaudhary, C. Ji, B. G. Thomas, S. P. Vanka, *Metall. Mater. Trans. B*, 42 (2011), 987–1007.
20. X. Miao, K. Timmel, D. Lucas, Z. Ren, S. Eckert, G. Gerbeth, *Metall. Mater. Trans. B*, 43(2012), 954-972.
21. R. Singh, B. G. Thomas, S. P. Vanka, *Metall. Mater. Trans. B*, 45 (2014), 1098-1115.
22. C. Kratzsch, K. Timmel, S. Eckert, R. Schwarze, *steel research int.*, 86 (2015), 400-410.

2016 EPD Congress

SYMPOSIUM: MATERIALS PROCESSING FUNDAMENTALS

Non-Ferrous Extractive Metallurgy

Antoine Allanore,
Massachusetts Institute of Technology

Guillaume Lambotte,
University of Massachusetts

EPD Congress 2016
Edited by: Antoine Allanore, Laura Bartlett, Cong Wang, Lifeng Zhang, and Jonghyun Lee
TMS (The Minerals, Metals & Materials Society), 2016

THE Cu-Ni-S SYSTEM AND ITS SIGNIFICANCE IN METALLURGICAL PROCESSES

Fiseha Tesfaye[1], Daniel Lindberg[1], Pekka Taskinen[2]

[1]Åbo Akademi University, Johan Gadolin Process Chemistry Centre, Piispankatu 8, FI-20500 Turku, Finland

[2]Aalto University School of Chemical Technology, Department of Materials Science and Engineering, Vuorimiehentie 2 K, FI-00076 Aalto, Finland

Keywords: Cu-Ni, Sulfide, Phase equilibria, Thermal stability, Phase transformation

Abstract

Due to increasing complexities of raw materials for base metals production the conventional pyrometallurgical processes are challenged. To make the appropriate modifications, thorough evaluations of the thermal stabilities of phases and phase assemblages which are commonly encountered in these processes are essential. In this work, phase relations and thermal stabilities of equilibrium phases in the Cu-Ni-S system have been reviewed. The calculated phase diagram of Cu-Ni system has been validated. At $T < 630$ K, in the Ni-rich corner, large scatter in data has been observed and discussed in detail. The other binary systems, Cu-S and Ni-S, have been also critically reviewed together with the ternary Cu-Ni-S system. The reviews were focused on compiling and discussing phase relations and thermal stabilities of the selected systems in the pyrometallurgical processes of Cu and Ni productions.

Introduction

Currently, complexities of the available raw materials are challenging the production of high grade metals by the conventional pyrometallurgical processes. Consequently, smelters are in need to modify their operating flow sheets and strategies for processing more complex feed materials economically, while meeting the strict environmental regulations. To make the appropriate modifications, accurate knowledge of existing phase relations and their thermal stabilities during the whole pyrometallurgical processes are essential.

Ore minerals for the production of non-ferrous metals constitute considerable amounts of Cu-Ni-sulfides. Some industrial mattes contain mainly Cu, Ni, S and Fe as well as traces of PGM metals. For instance, after conversion, average Cu-Ni-matte at PGM recovery plant contains Ni(50 wt.%), Cu(28 wt.%), S(21 wt.%) and Fe(3 wt.%) [1]. In the mattes, Cu-Ni-based alloys are the primary carriers of the PGMs. Processes for extracting valuable metals from secondary raw materials often involve dealing with Cu-Ni alloys. In applications other than extractive metallurgy, Cu-Ni alloys are among the most often used materials for metallization in electronic devices and are candidates for the new multicomponent solders [2]. Cupronickel (Cu-Ni) alloys are also used in variety of applications due to their high corrosion resistivity and thermal stability as well as working properties. Therefore, accurate knowledge of phase relations and thermal stabilities in the Cu-Ni-S system has considerable fundamental and practical interest in many aspects of extractive, chemical and physical metallurgy. This paper reviews and discusses phase relations and thermal stabilities in the Cu-Ni-S system with special emphasis placed on compiling and refining thermal stability data of phases and phase assemblages which are of interest in the pyrometallurgical processes of Cu and Ni productions.

The Cu-Ni System

The binary Cu–Ni system has been studied by several researchers [3-19]. The latest assessment was done by [3,4]. Contrary to the conflicting reports in the literature and an attempt to synthesize compounds at high pressure conditions, no intermediate or ordered Cu-Ni phases have been observed [7]. However, various alloying reactions between Cu and Ni were observed by many researchers [8-15].

The conventional Cu-Ni phase diagram is composed of the liquid phase and the fcc solid solution (s.s.). The solidus and liquidus boundaries have a characteristic lens shape with a narrow two-phase region, as shown in Figure 1. The solidus and liquidus boundaries are experimentally studied by [16-19]. The calculated solidus and liquidus lines agree well with these experimental data. The solid phase region constitutes a stable and wide miscibility gap, which is not well established. There are large deviations among the available experimental data [20-24] for the solid phase region. These large deviations could be due to the slow kinetics in the low-temperature region. Each of the available experimental data is superimposed on the calculated phase diagrams as shown in Figure 1. According to Turchanin et al. [3], at T = 298 K, the miscibility gap in the Cu-Ni system starts at ~5 at.% Ni. However, the recent first-principles studies of Teeriniemi et al. [25] suggested that the miscibility gap starts at 13 at.% Ni, at T=0 K.

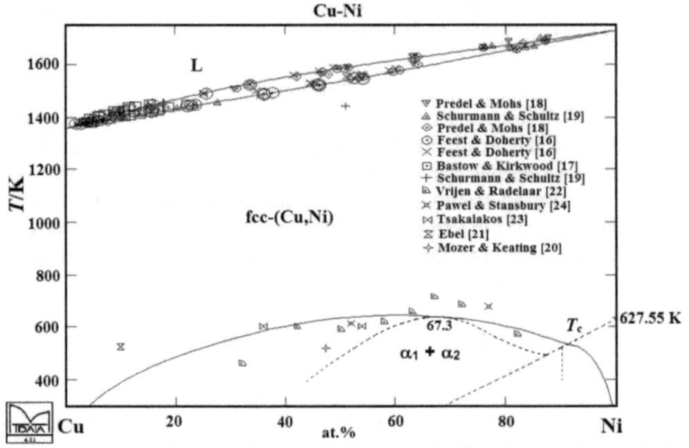

Figure 1. The calculated Cu-Ni phase diagram, with superimposed experimental data compiled in this study. The dashed line denotes the boundary of magnetic transformation.

In Figure 1, the parameter set with the magnetic contribution for the Cu-Ni system were taken from the MTOX database [5,6], version 7.0 without modifications. The thermodynamic interaction parameters of the liquid and fcc phases are based on the values reported in [4]. The SGTE Unary data [8,9] were used for the pure elements also without modifications.

Based on experimental data available in the literature, Chakrabarti et al. [7] made an assessment and suggested that at a critical temperature $T_c = 627.55$ K and composition of 67.3 at.% Ni, the fcc phase decomposes into two phases, α_1 and α_2. At $T < T_c$, α_2 changes from a paramagnetic to a ferromagnetic state. The suggested boundary of magnetic transformation line [7] is projected with dashed lines on the calculated phase diagram shown in Figure 1. The recent experimental study of Kravetsa et al. [26] on thin films of Cu-Ni alloys has also resulted in a similar low-temperature phase relations. However, owing to the differences in the thermodynamic properties of thin films and bulk materials [27], their determined curie temperatures are slightly higher. The latest study by Teeriniemi et al. [25], using the first-principles method, resulted in $T_c = 662$ K, at a composition of 78 at.% Ni, which is also higher than the previously suggested critical T_c and composition values of [7]. In contrast, Turchanin et al.'s [3] assessment shifts the critical point to lower values, $T_c = 606$ K and $x_{Ni} = 59$ at.%. Recently, Xiong et al. [28] recommended that the SGTE values of T_c and magnetic moment (ß) for fcc Ni needs to be revised; the β for fcc Ni seems to be underestimated, T_c overestimated. Lattice stability of pure Ni in the SGTE compilation also needs to be modified, which is a valid recommendation.

The Cu-S System

Cu-sulfides exceptionally frequently display rapid phase transformation at very low temperatures. At standard pressure conditions, 13 different binary Cu-S-compounds have been reported to be stable. A summary of all identified stable solid phases were compiled in Table I.

Table I. Minerals and phases in the Cu-S system [31-33].

Mineral name	Formula	Thermal stability (K)		Remarks
		Min.	Max.	
Chalcocite (mono)	Cu_2S	-	376	Inverts to hex form.
Chalcocite (hex)	Cu_2S	376	~708	Inverts to cubic form.
Chalcocite (cubic)	Cu_2S	~708	1402	Forms complete s.s. with $Cu_{1.8}S$.
Chalcocite (tetra)	Cu_2S	-	773	Stable only at $P > 1$ kbar.
Djurleite (ortho)	$Cu_{1.97}S$	-	366 ± 2	Decomposes.*
Digenite (cubic)	$Cu_{1.8}S$	348	356	Inverts to high-T cubic form.
High-Digenite (cubic)	$Cu_{1.8+x}S$	356	1402	Complete s.s.
Anilite (ortho)	$Cu_{1.75}S$	-	348 ± 3	Decomposes to $Cu_{1.8}S$ + CuS [34].
Yarrowite (hex-R)	$Cu_{1.125}S$	-	430	Thermodynamically stable?
Covellite (hex-R)	CuS	-	780	-
Roxbyite (mono)	$Cu_{1.78}S$	-	-	-
Geerite(cubic)	$Cu_{1.60}S$	-	-	-
Spionkopite (hex-R)	$Cu_{1.4}S$	-	-	-
Villamaninite(cubic)	CuS_2	-	-	-
-	CuS_5	-	823	High-P synthesis. Cubic?

* Decomposes into Cu_2S(hex) and Cu_2S-$Cu_{1.8}S$(cubic) phases.

Most of these phases are closely related both structurally and compositionally. For instance, the orthorhombic anilite transforms up on grinding to a metastable form of digenite and a small deficiency of Cu in Cu_2S results in a new phase $Cu_{1.97}S$ with a structure close to that of Cu_2S. Above $T = 773$ K the phase relations in the Cu-S system are straight forward; two intermediate phases, cubic solid solution series which ranges in composition from chalcocite to degenite and covellite, which is stable in the presence of its autogenous vapor pressure up to 780 K. The Cu-S phase diagrams below $T = 1100$ K have been published in [29]. Thermochemical aspects of the Cu-S system have been summarized by Barton et al. [30].

The liquid miscibility gap in the system Cu-Cu_2S has been studied intensively during the last hundred years. Recently, Forsström et al. [35] have experimentally investigated this system above 1373 K. Their results are in agreement with the calculated phase diagram by MTDATA – software using Mtox database [5], version 7.0, as shown in Figure 2. Below $T < 1673$ K, most of the experimental studies [35-48] are generally in agreement. These studies indicate that the liquid miscibility gap has almost vertical boundaries in the temperature range from 1423 K to 1723 K. However, experimental data of [36,37] suggest that the liquid miscibility gap to end at temperatures below $T = 1773$ K.

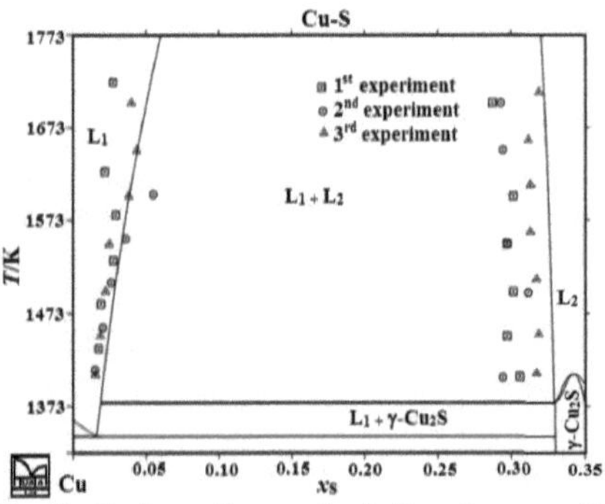

Figure 2. The calculated Cu-Cu_2S phase diagram including the open liquid miscibility gap, with superimposed experimental data of the latest study [35].

The Ni-S System

In contrast to the Cu-S system, the Ni-S system appears to have no multiple structures or phase transformations at low temperatures. Phase equilibria studies revealed phases with relatively high decomposition temperatures, rapid and reversible phase transitions. This system was extensively studied by [50-52], and, recently, modeled by [53]. The stable solid phases of the Ni-S system are; heazlewoodite (Ni_3S_2, hex), stable below $T = 829$K, $Ni_{3+x}S_2$, stable in the temperature range from 524 K to 1079 K, godlevskite (α-Ni_7S_6, ortho), stable below $T = 673$ K, Ni_7S_6, stable in the temperature range from 400 K to 846 K, millerite (NiS, hex), stable below $T = 652$ K, α-$Ni_{1-x}S$(hex), stable in the temperature range from 282 K to 1272 K, polydymite (Ni_3S_4, cubic), stable below $T = 629$ K, and vaesite (Ni_8S_2, cubic), stable below $T = 1280$ K [49]. For compositions with over 50 mol.% Ni, the liquid phase appears only above the eutectic temperature $T = 910$ K [53].

The Cu-Ni-S System

In order to improve metals extraction processes from mattes, the high-temperature phase relations and thermodynamic properties in the ternary Cu-Ni-S system have been studied by metallurgists intensively. This system has also been studied at elevated temperatures by geochemists, as a part of the more complicated Cu-Ni-Fe-S system, in attempts to gain knowledge concerning the conditions at which ores were deposited.

The only ternary phase in the Cu-Ni-S system, villamaninite $(Cu,Ni)S_2$, is stable at low temperatures and only in the absence of Fe [57]. Ryabko et al. [55] studied the Cu_2S-NiS system experimentally, including by the DTA method. Their proposed phase diagram is presented in Figure 3(a). The authors reported values for a ternary eutectic at 846.5 ± 1.5 K (21.5 wt.% S, 70.2 wt.% Ni and 8.8 wt.% Cu), and a peritectic at 1026 ± 3 K (14 wt.% Cu, 59.1 wt.% Ni, and 26.9 wt.% S). Phase relations in the Cu-Ni-S system in the temperature range from 373 K to 1473 K, including isothermal sections at 373, 473, 573, 673, 773 and 873 K, were described in [56,57]. A liquid phase emerges in equilibrium with $Ni_{3\pm x}S_2$, $Cu_{2\pm x}S$, and a Cu-Ni solid solution at an isothermal $T = 873$ K.

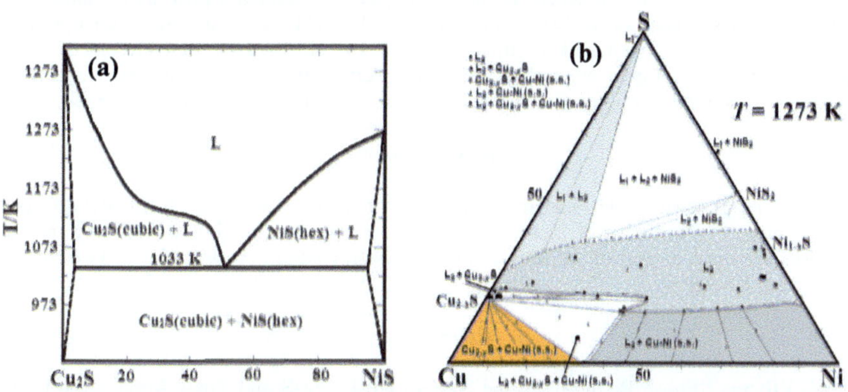

Figure 3. (a) Phase diagram of Cu_2S-NiS system, modified from [55], (b) isothermal section of the Cu-Ni-S system at 1273 K, modified from [54]. In both diagrams compositions are in wt.%.

Bruwer [54] experimentally studied the Cu-Ni-S system in the temperature range from 700 K to 1473 K. An isothermal phase diagram he proposed at 1273 K is shown in Figure 3(b). According to Bruwer [54], in the temperature range from 1373 K to 1473 K two immiscible liquid phases coexist in the Cu-rich compositions of the system. The [Cu-Ni(s.s.) + liquid] phase region at an isothermal $T = 1473$ K in the Ni-rich compositions extends toward higher Cu contents. At this isothermal condition, the [Cu-Ni(s.s.) + liquid] phase region boundary line intersects with the Cu-Ni join at ~80 wt.% Cu and intersects with the Ni-S join at ~23 wt.% Ni. At lower isothermal $T = 1273$ K, $Cu_{2-x}S$ and Cu-Ni(s.s.) phases coexist in the Cu-rich compositions at $x_S < 20$ wt.% S region. At the same isothermal condition $T = 1273$ K, Ni-rich Cu-Ni(s.s.) and a liquid phase co-exist with their phase region boundary line intersecting the Ni-S join at ~18 wt.% S. Above 50 wt.% Cu and below 20 wt.% S, three phases co-exist; [Cu-Ni(s.s.) + $Cu_{2-x}S$ + liquid], as can be seen in Figure 3(b). At higher sulfur contents (~20 wt.%), $Cu_{2-x}S$ and a liquid phase co-exist (see Figure 3(b)). At an isothermal $T = 973$ K, only a very small amount of liquid phase is left. At this lower isothermal condition $Cu_{2-x}S$, Cu-Ni(Ni-rich s.s.) and Ni_3S_2 still co-exist with a liquid phase.

The effect of small amount of Fe (from 1 wt.% to 10 wt.% Fe) in the Cu-Ni-S system has been experimentally investigated by [1]. The author concluded that in an isothermal condition increasing amount of Fe (up to 10 wt.%) favors formation of Ni-rich alloys. Effects of typical impurities such as As and Sb on the stabilities of this system have been reviewed elsewhere [58].

Summary and Conclusions

A review of phase relations and thermal stabilities in the Cu-Ni-S system has been conducted. The calculated phase diagram of Cu-Ni system has been validated with experimental data available in the literature. The Cu-Ni system is well known at high temperatures; however, in the solid phase region there are large deviations among the available experimental data. The solid immiscibility of Cu-Ni phase and the magnetic transformation appear to be not well stablished. In spite of some uncertainties concerning phase boundaries, phase relations in the binary Cu-S and Ni-S systems are well stablished. In the Cu-S system, at standard pressure conditions, 13 different binary Cu-S-compounds have been reported to be stable. The calculated Cu-Cu_2S phase diagram including the 'open ended' liquid immiscibility region has been presented. Phase relations in the Ni-S system are rather well stablished. Selected phase diagrams and high-temperature phase relations in the Cu-Ni-S system have been also reviewed and discussed.

Acknowledgements

The authors are grateful to the Academy of Finland for financial support. This work was made under the project "Chemistry of biomass impurities at reducing conditions in future thermal conversion concepts" as part of the activities of the Johan Gadolin Process Chemistry Center at Åbo Akademi University.

References

[1] W. Viljoen, "Phase Relations in the System Cu-Fe-Ni-S and Their Application to the Slow Cooling of PGE Matte" (Ph.D. thesis, University of Pretoria, 2001), 1-116.

[2] J. Miettinen et al., "Thermodynamic Description of the Cu–Ni–Pb System," *Cent. Eur. J. Chem.*, 9 (4) (2011), 743-749.

[3] M. Turchanin, P. Agraval, and A. Abdulov, "Phase Equilibria and Thermodynamics of Binary Copper Systems with 3d-metals. VI. Copper-Nickel System," *Powder Metall. Met. Ceram.*, 46 (9-10) (2007), 467-77.

[4] S. Mey, "Thermodynamic Re-evaluation of the Cu-Ni System," *CALPHAD*, 16 (1992), 255-260.

[5] J. Gisby, A. Dinsdale, I. Barton-Jones, A. Gibbon, P. Taskinen, "Predicting phase equilibria in oxide and sulphide systems" (EMC European Metallurgical Conference," June 11-14, Düsseldorf, Germany, 11-14 June 2007) 1721-1736.

[6] P. Taskinen, A. Dinsdale, J. Gisby, "Industrial Slag Chemistry: A Case Study of Computational Thermodynamics," *Scand. J. Met.*, 34 (2005), 100-107.

[7] D.J. Chakrabarti et al., "Cu-Ni (Copper-Nickel)," *Phase Diagrams of Binary Copper Alloys*, ed. P. Subramanian, D. Chakrabarti, and D. Laughlin (ASM International, Materials Park, OH, 1994), 276–286.

[8] F.J.A. Den Broeder, and S. Nakahara, "Diffusion-Induced Grain Boundary Migration and Recrystallization in the Cu-Ni System," *Scripta Metall.*, 17 (1983) 399-404.

[9] D. Liu, A. Miller, and K.T. Aust, "Diffusion Induced Grain Boundary Migration in Ni-Cu Diffusion Couples," *Acta Metall.*, 37 (1989), 3367.

[10] C.Y. Ma et al., "On the Kinetic Behavior and Driving Force of Diffusion Induced Grain Boundary Migration," *Acta Metall.*, 43 (8) (1995), 3113-3124.

[11] Y. Kawanami et al., "Growth Rate of Fine Grains Formed by Diffusion Induced Recrystallization in Ni Layer of Cu/Ni/Cu Diffusion Couples," *Mater. Trans. JIM*, 39 (1998), 218-224.

[12] Y. Yamamoto, S. Uemura, and M. Kajihara: *Proc. Int. Conf. Solid–Solid Phase Transformations* (Japan Inst. Met., Sendai, 1999), 593–598.

[13] Y. Yamamoto et al., "Observations on diffusion-induced recrystallization in binary Ni/Cu diffusion couples annealed at an intermediate temperature," *Mater. Sci. Eng.: A*, 312 (1-2) (2001), 176-186.

[14] Y. Yamamoto et al., "Kinetic features of Diffusion Induced Recrystallization in the Cu(Ni) System at 873 K," *Mater. Sci. Eng.: A*, 333 (1-2) (2002), 262-269.

[15] S.M. Schwarz et al., "Effects of Diffusion Induced Recrystallization on Volume Diffusion in the Copper-Nickel System," *Acta Mater.*, 51 (10) (2003), 2765-2776.

[16] E.A. Feest, R.D. Doherty, "The Cu–Ni Equilibrium Phase Diagram," *J. Inst. Metals*, 3 (1971), 102–103.

[17] B.D. Bastow, D.H. Kirkwood, "Solid/liquid equilibrium in the Copper–Nickel–Tin system determined by microprobe analysis." *J. Inst. Metals*, 99 (1971), 277–283.

[18] B. Predel, R. Mohs, "Thermodynamische Untersuchung flussiger Nickel–Kupfer Legierungen." *Arch Eisenhut*, 42 (1971), 575–579.

[19] E. Schurmann, and E. Schultz, "Untersuchengen zum Verlauf der Liquidus und Solidus linien in den Systemen Kupfer–Mangan und Kupfer–Nickel," *Z Metallkd.*, 62, (1971), 758–762.

[20] B. Mozer, D.T. Keating, S.C. Moss, "Neutron measurement of clustering in the alloy CuNi [copper–nickel]," *Physical Review B*, 175 (1968), 868–876.

[21] M.F. Ebel, "X-ray measurements on spinodal decomposition in copper-nickel alloys," *Physica. Status Solidi.* A, 5, (1971), 91–94.

[22] J. Vrijen, and Radelaar, "Clustering in copper–nickel alloys: a diffuse neutron-scattering studyd," *Physical Review B*, 17 (2) (1978) 409–421.

[23] T. Tsakalakos, "Spinodal decomposition in copper–nickel alloys by artificial composition modulation technique," *Scripta Metallurgica*, 15 (1981), 255–258.

[24] R.E. Pawel, and E.E. Stansbury, "The specific heat copper, nickel and copper–nickel alloys," *J. Phys. Chem. Sol.*, 26(1965), 607–613.

[25] J. Teeriniemi, P. Taskinen, K. Laasonen, "First-Principles Investigation of the Cu-Ni, Cu-Pd, and Ni-Pd Binary Alloy Systems," *Intermetallics* 57 (2015), 41-50

[26] A.F. Kravetsa et al., "Exchange-induced Phase Separation in Ni-Cu Films," *J. of Magnetism and Magnetic Materials*, (2012) 1-5.

[27] S. Wagner, A. Pundt, "Mechanical Stress Impact on Thin $Pd_{1-x}Fe_x$ Film Thermodynamic Properties," *Appl. Phys. Lett.*, 92, 051914 (2008) 1-3.

[28] W. Xiong, H. Zhang, L. Vitos, M. Selleby, "Magnetic phase diagram of the Fe-Ni system. *Acta Mater.* 59 (2011), 521-530.

[29] F. Tesfaye, and P. Taskinen, "Sulfide Mineralogy - Literature Review," (TKK-MT-214, ISBN 978-952-60-3271-9(printed), 2010) 59.

[30] P.B.Jr. Barton, "Solid Solutions in System Cu-Fe-S Part I. The Cu-S and CuFe-S Joins," *Econ. Geol.* 68. (1973), 455-465.

[31] D.J. Vaughan et al., eds., Reviews in Mineralogy and Geochemistry, vol. 61 (Concorde Pkwy Ste 500, USA: Mineralogical Society of America, 2006), 265-419.

[32] D.J. Chakrabarti, and D.E. Laughlin, "Cu-S (Copper-Sulpher)," *Binary alloy phase diagrams,* eds. T.B. Massalski, H. Okamoto, and L. Kacprzak (ASM, USA, 1990), 1467–1471.

[33] N. Morimoto, and G. Kullerud, "Polymorphism in Digenite," *Am. Mineral.* 48. (1963), 110-123.

[34] N. Morimoto, and K. Koto, "Phase Relations of the Cu-S System at Low Temperature: Stability of Anilite," *Am. Mineral.* 55. (1970), 106 – 117.

[35] A. Forsström, P. Taskinen, H. Johto,"Properties of the liquid miscibility gap in the system Cu-Cu_2S," (Aalto University Publication Series, SCIENCE and TECHNOLOGY 6/2012), 21.

[36] M. Moulki, and J. Osterwald, "Gleichgewichtsuntersuchungen am System Kupfer-Schwefel," (Dr. Ing. thesis, Technischen Universität Berlin, Germany, 1979) 80.

[37] K. Friedrich and M. Waehlert, "Untersuchungen über Schichten bildende Systeme II. Kupfer-Kupfersulfür," *Metal und Erz.*, 10 (1912/1913), 976-979.

[38] K. Bornemann, and K, Wagenman, "Die elektrische Leitfähigkeit der Metallegierungen im flüssigen Zustande," *Ferrum,* 11 (1914), 294-296.

[39] J. Niemelä, "Electrochemical measurement of sulfur activity in high temperatures," (Ph.D. thesis, Helsinki University of Technology, Finland, 1985) 60.

[40] M. Kitano, "The Diagram of the Cu-S-O System on Fundamental Reactions in the Copper Smelting Process," *Nippon Kinzoku Gakkaishi*, 28 (1964), 856-861.

[41] F. Johannsen, and H. Vollmer, "Untersuchungen im System Kupfer-Kupfersulfur," *Erzmetall*, 13 (1960), 313-322.

[42] R. Schuhmann, and O.W. Moles, "Sulfur Activities in Liquid Copper Sulphides," *Trans. AIME,* 191 (1951), 235-241.

[43] C.W. Bale, and J.M. Toguri, "Thermodynamics of the Cu-S, Fe-S and Cu-Fe-S systems," *Canadian Metallurgical Quarterly*, 15 (1976), 305-318.

[44] J. Koh, and A. Yazawa, "Thermodynamic Properties of the Cu-S, Fe-S and Cu-Fe-S Systems," *Bulletin of the Research Institute of Mineral Dressing and Metallurgy Tohoku University*, 38 (1982), 107-118.

[45] T. Yagihashi, "Measurement of the Equilibrium Relation between Sulfur in Molten Copper and H_2S-H_2 Gas Mixtures," *Nippon Kinzoku Gakkaishi*, 17 (1954), 483-487.

[46] V.P. Judin, and E. Eerola, "Thermodynamics of metallic impurities in Copper-saturated Copper Sulphide melts: I. The miscibility gap in the system Cu-Bi-S at 1423-1573 K," *Scandinavian Journal of Metallurgy*, 8 (1979), 128-132.

[47] J. Schmiedl et al, "Equilibrium studies in the system Cu-S-O," (Institution of Mining and Metallurgy, 86, 1977), C88-C91.

[48] D.J. Chakrabarti, and D.E. Laughlin, "The Cu-S System," (Bulletin of Alloy Phase Diagrams, 4, 1983).

[49] R.G. Arnold, and O.P. Malik, "The NiS-S System Above 980 °C – A Revision," *Econ. Geol.* (in press). 1974.

[50] R.G. Arnold, and G. Kullerud, "The NiS-NiS_2 Join" (Year Book I Carnegie Institution of Washington. Washington, 1956), 178-179.

[51] G. Kullerud, and R.A. Yund, "The Ni-S system and related minerals," *J. of Petrology,* 3 (1962) 126-175.

[52] E.F. Sinyakova, and V.I. Kosyakov, and V.A. Shestakov, "Investigation of the Surface of the liquidus of the Fe-Ni-S System at $x_s < 0.51$." *Metallurgical and Materials Transactions B*, 30B (1999), 715 722.

[53] P. Waldner, and A.D. Pelton, "Thermodynamic modeling of the Ni–S system," *Zeitschrift fü rMetallkunde*, 95 (2004), 672-681.

[54] J.S. Bruwer, "Experimental investigation of the phase relations in the system Cu-Ni-S in the temperature range 1200 °C to 700 °C. University of Pretoria; unpublished thesis, (1996), 157.

[55] A.G. Ryabko, G.I. Grodinskii, and V.F. Serebryakov, *Izv. Vyssh. Uchebn. Zaved., Tsvetn. Metall.*, 4 (1980), 23-26.

[56] G. Kullerud, R.A. Yund, and G.H. Moh, "*Phase Relations in the Cu-Fe-S, Cu-Ni-S, and Fe-Ni-S Systems*", in: Magmatic Ore Deposits, ed. H.D.B Whson,. (Economic Geology Monograph 4, 1969), 343.

[57] G.H. Moh, and G. Kullerud, "The Cu-Ni-S System and Low Temperature Mineral Assemblage", *Ore Genesis*, 2 (182), 677-688.

[58] F. Tesfaye, and P. Taskinen, "Thermodynamics and Phase Equilibria in the (Ni, Cu, Zn)-(As, Sb, Bi)-S Systems at Elevated Temperatures (300–900 °C)" (TKK-MT-216. ISSN 1455-2329 (printed), 2011). DOI: http://dx.doi.org/10.13140/2.1.4901.2803

[illegible]

[illegible]

[illegible]

[50] [illegible] and [illegible] (eds.), "The [illegible]," New Book [illegible] on Stainless [illegible] (1970) [illegible].

[illegible] and [illegible] system and related minerals" [illegible] (1988) [illegible].

[illegible] S.R. [illegible] and V. A. [illegible], "[illegible] of the [illegible] in the [illegible] System [illegible]," [illegible] B, [illegible] (1999) [illegible].

[53] P. Waldner and A.D. Pelton, "Thermodynamic modeling of the Ni-S system," Zeitschrift für Metallkunde, 92 (2001) 672-681.

[54] [illegible], "[illegible] and [illegible] of the phase relations in the system Cu-Ni-S in the temperature range 200 °C [illegible]" [illegible] 1987.

[illegible]

[illegible]

[illegible] "The Cu-Ni-S System and [illegible]" [illegible]

[illegible] (Ag, Sb, Bi, S) [illegible] at Elevated Temperatures (500–800 °C) [illegible] 2011 [illegible] http://dx.doi.org/10.1[illegible]

2016 EPD Congress

SYMPOSIUM: MATERIALS
PROCESSING FUNDAMENTALS

Iron and Steelmaking - Thermodynamic, Reduction and Physical Metallurgy

Laura Bartlett,
Texas State University

Lifeng Zhang,
University of Science and Technology
Bei ing

EPD Congress 2016
Edited by: Antoine Allanore, Laura Bartlett, Cong Wang, Lifeng Zhang, and Jonghyun Lee
TMS (The Minerals, Metals & Materials Society), 2016

REDUCTION INETICS OF MAGNETITE CONCENTRATE PARTICLES WITH HYDROGEN AT 1150 1600 °C RELEVANT TO A NOVEL FLASH IRONMA ING PROCESS

Mohamed Elzohiery[1], Yousef Mohassab[1], Amr Abdelghany[1], Shengqin Zhang[1,3], Feng Chen[1,2], and Hong Yong Sohn[1]

[1] Department of Metallurgical Engineering, University of Utah, Salt Lake City, Utah 84112, USA
[2] School of Minerals Processing and Bioengineering, Central South University, Changsha, Hunan 410083, China
[3] School of Metallurgy and Materials Engineering, Chongqing University of Science and Technology, Chongqing 401331, China

Keywords: Reduction Kinetics, Magnetite Concentrate, Hydrogen Reduction, Ironmaking, Flash Reduction

Abstract

A novel ironmaking process is under development at the University of Utah aimed at producing iron directly from iron oxide concentrate in a flash reactor. This process will reduce hazardous emissions and save energy. The kinetics of magnetite reduction with hydrogen was previously investigated in our laboratory in the temperature range 1150 to 1400 °C at large temperature increments (~100 °C increments). Due to the significant melting that occurs above 1350 °C, the reduction kinetics was measured and analyzed in two distinct temperature ranges of 1150 to 1350 °C and 1350 to 1600 °C (~50 °C increments). Experiments were performed using magnetite concentrate particles of different sizes under various hydrogen partial pressures and residence times. Reduction degrees of more than 90 % were achieved in a few seconds at temperatures as low as 1250 °C. Different rate expressions were needed to obtain reliable agreement with experimental data.

Introduction

Ironmaking by the blast furnace (BF) is a multi-step process which consumes energy and produces hazardous emissions during its steps as sintering/pelletization, coke-making, etc. The novel ironmaking process being developed at the University of Utah [1-10] aims at producing iron directly from iron oxides concentrate particles in a gas-solid flash reaction utilizing hydrogen gas or natural gas as the reductant and fuel within temperature range 1200-1600 °C. Without the need of pelletization, sintering or coke, this process significantly saves energy and cuts CO_2 emissions. For designing flash reactors, it is important to determine the reduction rate equations. Natural gas will be used in this process by partial oxidation with oxygen in a flame where H_2 and CO gas are produced as a results of this in situ reforming. Although in general the reduction kinetics by H_2 is much faster than that by CO, it is necessary to study the reduction kinetics of each gas separately in order to analyze the complex kinetics of magnetite reduction by H_2 + CO mixtures.
The kinetics of magnetite reduction by hydrogen gas was studied in this project previously [11]. In that work, the kinetics model formulated using various runs at temperatures 1150, 1200, 1300,

and 1400 °C. Due to their gangue content, the concentrate particles melt at temperatures higher than 1350 °C, which will change the reduction mechanism. It is important to study the reduction kinetics in the temperature range lower than 1350 °C separately from that at higher temperatures using smaller temperature increments (50 °C). In this work, two rate equations were formulated for two temperature ranges: 1150-1350 °C and 1400-1600 °C.

Experimental Work

Magnetite concentrate particles were used in this work with 3 size fractions: 20-25, 32-38 and 45-53 μm. The concentrate particles are irregularly shaped and angular. As the concentrate particles experience melting at temperature higher than 1350 °C, the mechanism of reduction is expected to change together with the reduction kinetics. Therefore, the temperature range was divided into 1150-1350 °C and 1400-1600 °C. For each temperature range, a reduction rate equation was formulated including the effect of particle size, reducing gas partial pressure, and temperature.

Two drop tube reactors (DTR) were used for measuring reduction rate of concentrate particles. Both reactors had a dilute particles-gas conveyed system in which the fine particles fall in the tube with the reducing gas. The reactor system consists of a vertical tubular furnace housing an alumina tube, a pneumatic powder feeder, gas delivery lines, a powder cooling and collecting system, and an off-gas outlet.

For nominal particle residence time calculation, temperature profile is performed before each temperature set in order to define the reaction zone inside the reactor as ($T_s \pm 30$ °C) where T_s is the set temperature. A honeycomb was placed at the beginning of the reaction zone in order to straighten the flow inside the reactor. Concentrate was fed through a water-cooled tube at the beginning of the reaction zone directly. The reaction zone was corrected according to the method described by Chen et al. [9, 10].

During heating up the reactor, nitrogen gas was flowed until the target temperature was reached then the reducing gas was introduced according to the experiment conditions. Nitrogen gas was added to adjust the partial pressure of the hydrogen and the nominal particle residence time. Gases flow rates were controlled using rotameter flow meters.

Once all the gases are introduced and the reactor temperature is stable, magnetite concentrate is fed into the reactor at the specified feeding rate. All the experimental conditions are designed to maintain the excess hydrogen gas of more than 500% to ensure that the gaseous reactant concentration remains essentially unchanged over the entire reactor length for accurate quantification of the effect of partial pressure. The reduced powder was collected in the collection bin after the experiment. The collection bin was kept hot during the experiment to prevent any water condensation in the sample collection bin then after the experiment, the collection bin was cooled in water under N_2 atmosphere to avoid any re-oxidation of the product.

Samples collected were analyzed by ICP-OES to determine the iron fraction which was used to calculate the reduction degree. SEM analysis was employed to study morphology of the reduced samples.

The relevant equations for calculating the nominal particles residence time, %excess hydrogen, excess driving force (EDF), and reduction degree (RD%) was described elsewhere [9].

Reproducibility of the experiments was confirmed by the consistency of the results of repeated experiments under the same conditions including the reproducibility of the analysis method. The experimental precision was within ± 5%.

Results

Reduction Kinetics at Temperature Range 1150-1350 °C

The samples produced by the reduction experiments were analyzed using ICP-OES. Then, the reduction degree (%) change with nominal particle residence time (s), an example is shown in Figure 1. More than 90% reduction degree was achieved at 1250 °C with H_2 partial pressure 0.3 atm and particle residence time 3.6 s. The kinetics of hydrogen reduction of hematite concentrate particles was studied previously by Chen et al. [9] and the nucleation and growth kinetics model was found to best describe the reduction kinetics. Based on their studies, the nucleation and growth kinetics expression given by the following equation was tested and was found to describe the reduction rate of magnetite concentrate particles:

$$(-Ln(1\text{-}X))^{1/n} = k_{app}t \quad (1)$$

$$(-Ln(1\text{-}X))^{1/n} = k' \times e^{\frac{E_a}{RT}} \times f(p_{H_2}) \times h(D_p) \times t \quad (2)$$

where X is the iron fraction, n is the Avrami parameter, k_{app} is the apparent rate constant, k' is the pre-exponential factor, E_a is the activation energy for the reaction, $f(p_{H_2})$ is the function for the hydrogen partial pressure dependence, $h(D_p)$ is the particles size dependence function and t is the nominal particle residence time.

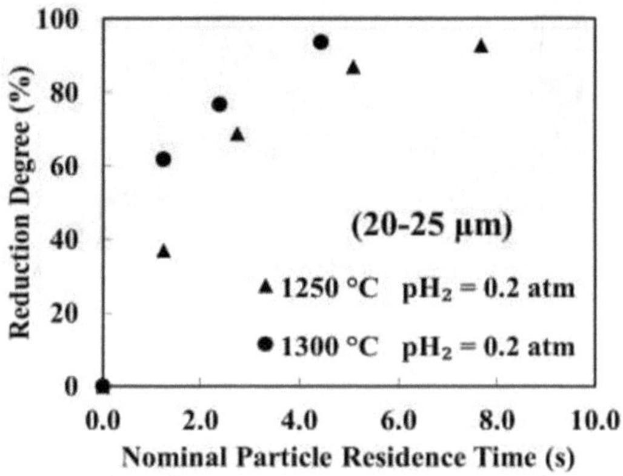

Figure 1. Change of reduction degree (%) with the nominal particle residence time at different conditions using particles size (20-25 microns).

The nucleation and growth mechanism was evident as seen in the SEM micrographs, Figure 2, where nuclei were formed on active sites on the particles. Avrami parameter, n, was obtained by plotting $Ln(-Ln(1\text{-}X))$ with $Ln(t)$ and found to be 1, as shown in Figure 3. The reduction kinetics had a first-order dependence on the hydrogen partial pressure, as shown in Figure 4. When studying the particle size effect, it was found that there is no noticeable effect on the reduction kinetics. The activation energy was calculated according to the Arrhenius equation to be 190 kJ/mol, as shown in Figure 5. As shown by XRD and SEM previously [9], a combination of different iron oxides and iron metal is present at any given moment during the reduction process. Therefore, the rate of reduction in this work was represented by the overall fraction of oxygen removed without distinguishing the formation of different iron oxide phases. In this global equation, the rate of oxygen removal from the iron oxide without specifying a reduction reaction for an oxide phase. The complete rate equation was:

$$(-Ln(1-X)) = 6.41\times10^{6} \times e^{\left(\frac{-190{,}000}{RT}\right)} \times \left[(p_{H_2}) - \left(\frac{p_{H_2O}}{K_{eq}}\right)\right]_{lm} \times t \tag{3}$$

where R is 8.314 J/mol K, T is in K, p is in atm, and t is in seconds. It was found that Eq. (3) can satisfactorily predict the reduction degree as shown in Figure 6.

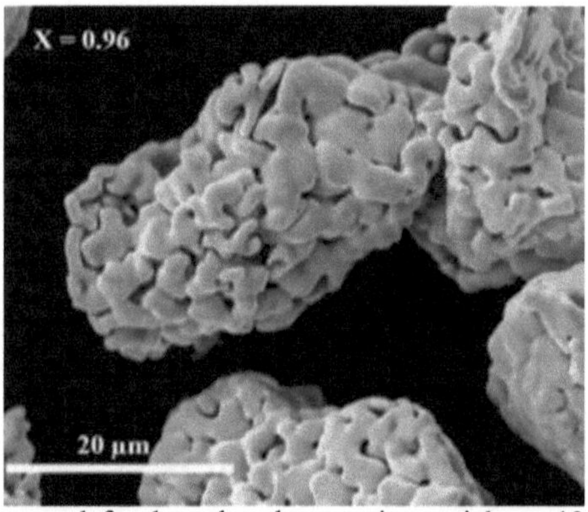

Figure 2. SEM micrograph for the reduced magnetite particles at 1250 °C and X = 0.96.

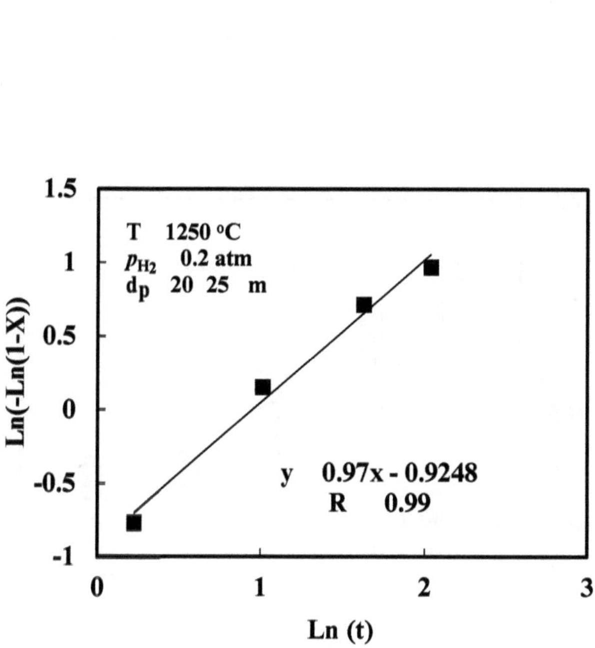

Figure 3. Determination of the Avrami parameter.

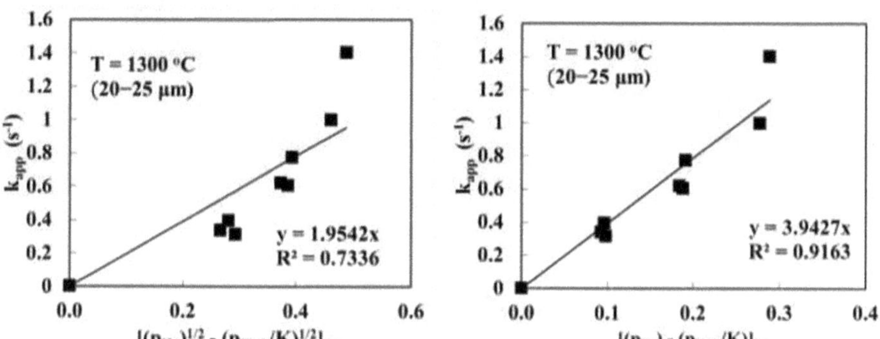

Figure 4. Obtaining the pH_2 dependence.

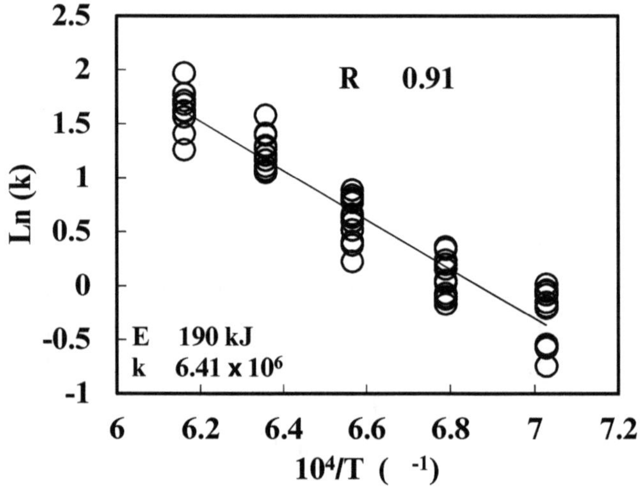

Figure 5. The Arrhenius plot for obtaining the activation energy and the rate constant for the reaction.

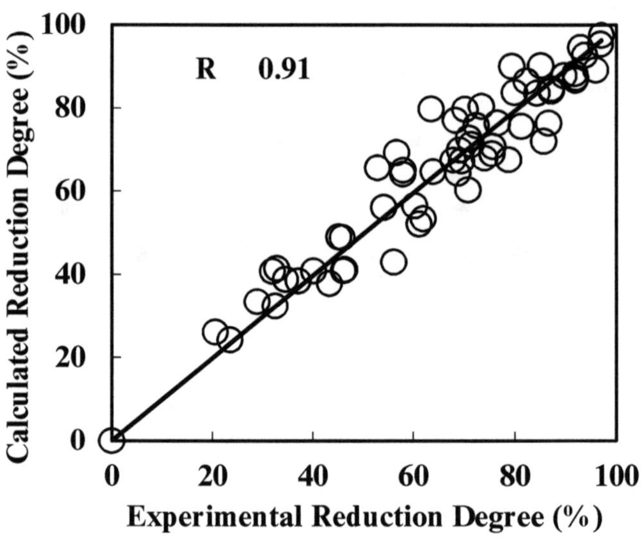

Figure 6. Calculated reduction degree (%) compared to the experimental reduction degree (%).

Reduction Kinetics at Temperature Range 1400-1600 °C

At 1400 °C, reduction degree of 95% was achieved at H_2 partial pressure 0.15 and particles residence time 1.8 s, as shown in Figure 7. Nucleation and growth kinetics expression was tested and found to describe the reduction kinetics in this temperature range, with an Avrami parameter n = 1. The reducing gas partial dependence was found to be of the first order. The particle size dependence was determined and the activation energy was found to be 182 kJ. The complete rate equation is:

$$(-Ln(1-X)) = 6.31\times10^{7}\, e^{(\frac{-182000}{RT})}\, (D_p)^{-1}\, [(p_{H_2}) - (\frac{p_{H_2O}}{K_{eq}})]_{lm}\; t \qquad (4)$$

where R is 8.314 J/mol K, T is in K, p is in atm, and t is in seconds. The calculated versus the experimental results are shown in Figure 8 with a good agreement.
Although the activation energy in the two temperature ranges were close, the pre-exponential factors were an order of magnitude different. In addition the effect of particle size in the temperature range 1400-1600 °C was observed, perhaps because of the melting of the particles which removes any cracks and porosity.

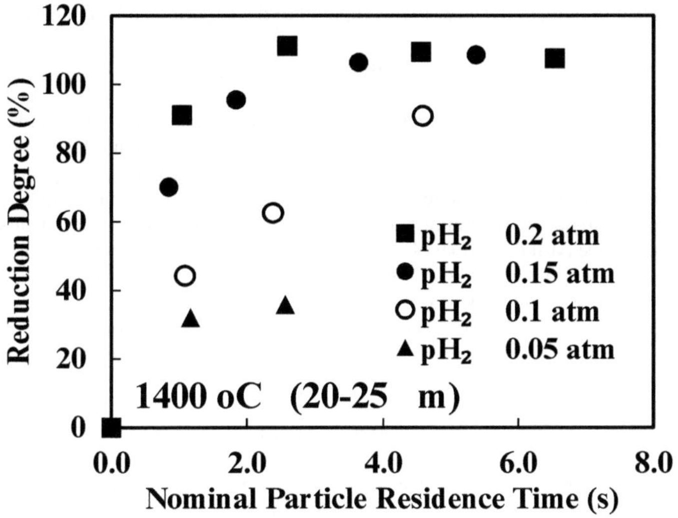

Figure 7. Change of reduction degree (%) with the nominal particle residence time at 1400 °C and different H_2 partial pressures using particles size (20-25 microns).

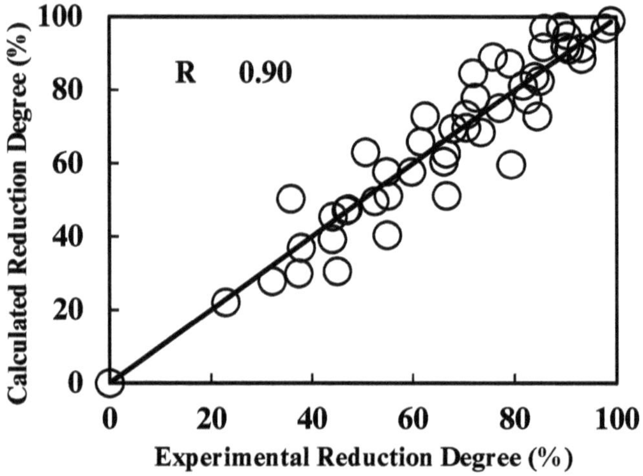

Figure 8. Calculated reduction degree (%) compared to the experimental reduction degree (%).

Conclusions

The kinetics of magnetite concentrate reduction with hydrogen gas was studied in the temperature ranges 1150-1350 °C and 1400-1600 °C. It was found that magnetite particles can be reduced to more than 90% in a few seconds available in the gas-solid flash reactor at temperature as low as 1250 °C. It was found that the nucleation and growth kinetics describe the reduction in the two temperature ranges. In both temperature ranges, an Avrami parameter $n = 1$ and 1st-order reduction rate dependence on the partial pressure of hydrogen were observed. In the temperature range 1150-1350 °C, an activation energy value of 190 kJ/mol was determined whereas in the temperature range 1400-1600 °C, 182 kJ/mol was obtained. In addition, effect of particle size was present in the temperature range 1400-1600 °C. Two complete rate equations were obtained representing the reaction rates at each temperature range. The rate equations showed good agreement with the experimental results with magnetite concentrate particles.

Acknowledgments

The authors thank Andrew Laroche, Omar Kergaye, Caio Melo, and Jacqueline De Oliveira Cota for the help with the experimental runs and the analytical work using ICP. The authors acknowledge the financial support from the U.S. Department of Energy under Award Number DEEE0005751 with cost share by the American Iron and Steel Institute (AISI) and the University of Utah.

employees, makes any warranty, express or implied, or assumes any legal liability or responsibility for the accuracy, completeness, or usefulness of any information, apparatus, product, or process disclosed, or represents that its use would not infringe privately owned rights. Reference herein to any specific commercial product, process, or service by trade name, trademark, manufacturer, or otherwise does not necessarily constitute or imply its endorsement, recommendation, or favoring by the United States Government or any agency thereof. The views and opinions of authors expressed herein do not necessarily state or reflect those of the United States Government or any agency thereof.

References

1. M.Y. Mohassab-Ahmed and H. Y. Sohn, *Method and Device for Digestion of Materials in a Microwave Oven*, 2012, US Patent App. 61/651.
2. M.Y. Mohassab-Ahmed, H. Y. Sohn, "Effect of Water Vapor Content in H_2–H_2O–CO–CO_2 Mixtures on the Equilibrium Distribution of Manganese between CaO–MgO_{sat}–SiO_2–Al_2O_3–FeO–P_2O_5 Slag and Molten Iron." *Steel Res. Int.*, 85 (2014), 875-884.
3. M.Y. Mohassab Ahmed, "*Phase Equilibria between Iron and Slag in CO/CO_2/H_2/H_2O Atmospheres Relevant to a Novel Flash Ironmaking Technology*" (PhD Dissertation, The University of Utah, 2013).
4. M.Y. Mohassab-Ahmed, H. Y. Sohn, "Effect of Water Vapor Content in H_2-H_2O-CO-CO_2 Mixtures on the Activity of Iron Oxide in Slags Relevant to a Novel Flash Ironmaking Technology." *Ironmaking Steelmaking*, 41(2014), 665 - 675.
5. Y. Mohassab and H. Sohn, "Effect of Water Vapour on Distribution of Phosphorus between Liquid Iron and MgO Saturated Slag Relevant to Flash Ironmaking Technology." *Ironmaking Steelmaking*, 41 (2014), 575-582.
6. Y. Mohassab and H.Y. Sohn, "Effect of Water Vapor on Sulfur Distribution between Liquid Fe and MgO!Saturated Slag Relevant to a Flash Ironmaking Technology." *Steel Res. Int.*, 86 (2014), 753–759.
7. Y. Mohassab and H.Y. Sohn, "Analysis of Slag Chemistry by FTIR!RAS and Raman Spectroscopy: Effect of Water Vapor Content in H_2-H_2O-CO-CO_2 Mixtures Relevant to a Novel Green Ironmaking Technology." *Steel Res. Int.*, 86 (2014), 740-752.
8. M.Y. Mohassab-Ahmed, H.Y. Sohn, and L. Zhu, "Effect of Water Vapour Content in H_2-H_2O-CO-CO_2 Mixtures on MgO Solubility in Slag under Conditions of Novel Flash Ironmaking Technology." *Ironmaking Steelmaking*, 41 (2014), 575-582.
9. F. Chen, Y. Mohassab, T. Jiang, and H.Y. Sohn, "Hydrogen Reduction Kinetics of Hematite Concentrate Particles Relevant to a Novel Flash Ironmaking Process." *Metall. Mater. Trans. B*, 46 (2015), 1133-1145.
10. F. Chen, Y. Mohassab, S. Zhang, and H.Y. Sohn, "Kinetics of the Reduction of Hematite Concentrate Particles by Carbon Monoxide Relevant to a Novel Flash Ironmaking Process." *Metall. Mater. Trans. B*, 46 (2015), 1716-1728.
11. H. Wang and H.Y. Sohn, "Hydrogen Reduction Kinetics of Magnetite Concentrate Particles Relevant to a Novel Flash Ironmaking Process." *Metall. Mater. Trans. B*, 44 (2012), 133-145.

EPD Congress 2016
Edited by: Antoine Allanore, Laura Bartlett, Cong Wang, Lifeng Zhang, and Jonghyun Lee
TMS (The Minerals, Metals & Materials Society), 2016

HYDROGEN REDUCTION INETICS OF MECHANICALLY ACTIVATED MAGNETITE CONCENTRATE

Ricardo Morales-Estrella[1], Juan Ruiz-Ornelas[1], Yousef Mohassab[2], Noemi Ortiz-Lara[1], and Hong Yong Sohn[2]

[1] Instituto de Investigación en Metalurgia y Materiales, Universidad Michoacana de San Nicolás de Hidalgo, Morelia, Michoacán, C.P. 58030, México.
[2] Department of Metallurgical Engineering, University of Utah, Salt Lake City, UT 84112, USA.

Keywords: Hydrogen reduction, magnetite, mechanical activation, reduction kinetics.

Abstract

The effect of mechanical activation on the reduction kinetics of magnetite concentrate by hydrogen was studied. The magnetite concentrate was milled for 8 h using a planetary mill. After the milling process, the average particle size was reduced from 14 to 4.4 μm resulting in a lattice microstrain of 0.30. Thermogravimetric experiments were conducted to focus on the chemical reaction as the rate controlling factor by eliminating external mass transfer effects and using a thin layer of particles to remove interstitial diffusion resistance. The onset temperature of reduction was decreased due to the mechanical activation, and the degree of conversion was decreased by sintering of particles which was confirmed by SEM analyses. In view of the results, a reaction rate expression is discussed from which the activation energy is calculated.

Introduction

The gaseous reduction of iron oxides is a complex heterogeneous reaction in which the intrinsic rate mechanisms depend upon many parameters such as temperature, type of gaseous reactant, and particle size. There is, therefore, extensive research focused on different oxides reduced under different conditions which often lead to discrepancies among published work. Pineau *et al.*[1] studied the reduction of synthetic magnetite by hydrogen in the temperature range of 483 - 1223 K. They used about 100 mg of particles with 10 - 20 μm in diameter. The Arrhenius plot revealed three distinct slopes from which they obtained different apparent activation energies. That is, 200 kJ/mol, 71 kJ/mol and 44 kJ/mol at T < 523 K, 523 K < T < 663 K and T > 663 K, respectively. Pourghahramani and Forssberg [2] investigated the effect of mechanical activation on the hydrogen reduction of high purity hematite using 95 mg of sample and a hydrogen flow rate of 100 mL/min. They reported a decrease of 90 K in the onset of the reducing temperature with respect to the non-activated sample. In addition, wüstite phase was not detected as an intermediate phase during nonisothermal experiments with heating rates of 10 to 15 K/min. Under thermodynamic equilibrium, wüstite is an unstable phase below 843 K; however, as shown in Ref. [3], wüstite could be an intermediate phase during the reduction of hematite by hydrogen below 843 K under irreversible thermodynamic conditions. Research of industrial interest on iron oxides is mostly carried out in the form of pellets or sinter. Recently, Wang and Sohn [4] investigated the reduction kinetics of fine magnetite concentrates using a novel process for rapid reduction; they developed a rate expression for the reduction process and obtained an activation energy value of 463 kJ/mol in the temperature range of 1243 - 1673 K. The objective of this work was to study the reduction kinetics of mechanically activated magnetite concentrates by hydrogen. The experimental conditions were set up to obtain the rate of the chemical reaction as the rate controlling factor. The results were compared to non-activated samples and a rate expression was established for the reduction of both activated and non-activated samples.

Experimental

Materials

Magnetite concentrate from the Mesabi Range of the U.S. was used as the solid reactant in this study and it is composed of 0.688, 0.041, and 0.271 of mass fractions of total iron, gangue, and oxygen, respectively. Argon and hydrogen gases, with 99.999% purity, were used in the thermogravimetric experiments.

Mechanical Activation

Mechanical activation of magnetite concentrate was induced by milling in a planetary ball mill (Restch, PM400) at 300 rpm. Stainless steel balls of 10 mm in diameter were used as the grinding media; the ball-to-powder weight ratio was kept at 10:1.

Characterization

The particle size distribution of as-received and milled particles was obtained by Dynamic Light Scattering (DLS) using a Coulter LS 100Q apparatus. The morphology of the samples was characterized using a JEOL 6400 scanning electron microscope (SEM) with a NORAN-EDS attachment. X-ray analyses were performed using a Bruker D8 Advance powder diffractometer with Bragg-Brentano geometry. The Bragg peaks of samples were recorded at 40 mA, 40 kV with a step size of 0.02 degree and 5 s per step using Cu-Kα radiation and a nickel filter to suppress fluorescence. The lattice strain of milled samples was calculated using TOPAS software provided by Bruker Corporation.

Isothermal reduction was carried out using a vertical thermogravimetric unit (Setaram, Setsys Evolution 16/18, France) having a detection limit of 0.03 μg. The balance is controlled by a PC through an interface. An Al_2O_3 crucible, with 8 mm inner diameter and 1 mm inner height, was used to hold the magnetite concentrate. To avoid particle agglomeration, a drop of alcohol was used to evenly disperse the sample on the bottom surface of the crucible. The temperature of the furnace was controlled by a Pt-Pt/13% Rh (S-type) thermocouple placed just below the crucible. The themogravimetric cycle began by heating the sample up to the reducing temperature at a heating rate of 20 K/min under helium gas with a constant flow of 15 mL/min. When the sample temperature was stabilized, the helium gas flow was stopped and hydrogen gas was introduced and kept at a constant flow until the sample weight loss remained unchanged. The mass change during TGA experiments was recorded at 1 s intervals. All experiments were corrected against the noise of the hydrogen gas velocity on the sample by running blank experiments under identical conditions. Preliminary reduction experiments were performed to establish the optimal parameters to obtain the rate of the chemical reaction as the rate controlling factor.

Results and Discussion

Magnetite concentrates were subjected to different milling times to both decrease particle size and introduce micro strain. Thus, Figure 1 shows that after two hours of milling time, the mean particle size seems to remain unchanged. It can be expected that after the initial particles fracture into smallest particles, the repeated mechanical deformations caused by high energetic ball–powder–ball and ball–powder–chamber collisions would introduce micro strain into the lattice. The micro strain for samples milled at different times was calculated using TOPAS software and is shown in Figure 2. From the above results, the sample milled for 8 hours was used to study the effect of mechanical activation on the reduction kinetics. For direct comparison, the as-received concentrate was also studied under similar reducing conditions.

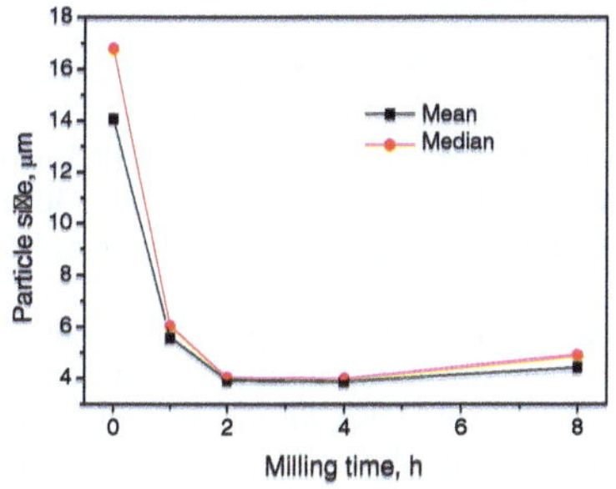

Figure 1. Particle size of magnetite concentrates as function of milling time.

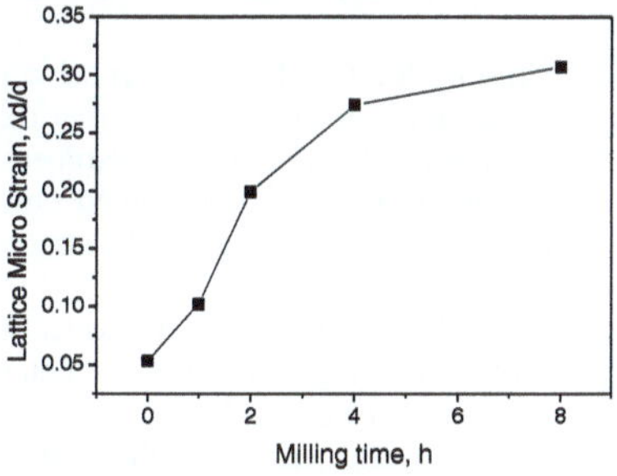

Figure 2. Lattice micro strain as function of milling time for the magnetite concentrate used in this work.

The particle size distribution of the two concentrate powders used in this work was measured and the results are shown in Figure 3. It can be seen that the milling process not only decreases the particle size but also narrows its distribution to a significant extent. The mathematical models for heterogeneous gas-solid reactions describe that the reaction rate depends on the radius of the spherical particle and therefore a constant particle size in the system is usually assumed. R. Morales *et al.* [5] took into account the range of particle size distribution in a rate expression for the chemical reaction showing negligible impact on the calculation of the activation energy.

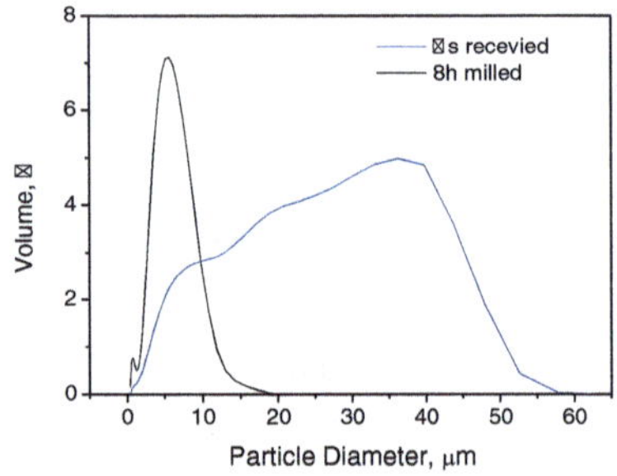

Figure 3. Particle volume distribution for magnetite concentrates.

It has been reported that the mechanical activation of hematite intensifies its thermal reactivity with hydrogen gas [2]. To verify whether the milled magnetite concentrate is able to decrease its onset temperature of reduction, nonisothermal experiments were performed for milled and as-received samples under predetermined conditions. The first derivative of each nonisothermal reduction curve was plotted against temperature in Figure 4 to show the onset temperature of reduction.

Figure 4 a) depicts earlier stages of reduction where the reactivity of the concentrate increases as the milling time is increased. It is to be noted that all curves show a discontinuity between 650 K and 665 K which correspond to $X < 0.1$. This behavior indicates that an intermediate phase is likely to be occurring. Although the Fe-O equilibrium diagram shows that wüstite is stable at temperatures above 843 K, the current experimental set up allows a constant removal of oxygen from the system leading to a nonequilibrium process in which case the wüstite phase could be present as a transitory phase. Note that the fraction of reduction at which the discontinuity occurs is lower than the theoretical value of X for the reduction of magnetite to wüstite, namely, 0.25. However, during the nonisothermal process, as temperature increases, the unstable wüstite phase is likely to get reduced to metallic iron before the reduction of magnetite to wüstite is completed. Figure 4 b) shows the whole range of the reduction fraction; it is observed that at about 730 K (457 °C with $0.22 < X < 0.28$), the derivative tends to increase as the milling time is increased. Thereafter, at about 820 K (547 °C with $0.73 < X < 0.80$), the derivative value decreases sharply. This sudden decrease indicates a mass transfer aspect affecting the reaction rate which is likely to be triggered by the relatively large amount of sample used in addition with the late stage of reduction. It is more likely that the mechanical activation has a greater impact on the initial stage of reduction. It is thought that the energy supply by the milling would form a metastable magnetite phase with appreciable structural defects.

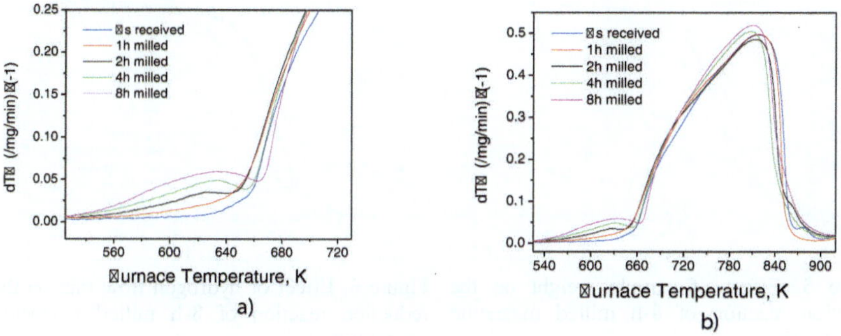

Figure 4. First derivative of the nonisothermal reduction experiments; a) derivative curves correspond to $0 < X < 0.25$, b) derivative curves correspond to $0 < X < 1$,

As mentioned earlier, the results shown in Figures 4 suggest a two-step reaction. This mechanism is currently being confirmed by the SEM and XRD analyses on quenched samples. Hence, the chemical reaction for the reduction of Fe_3O_4 by hydrogen gas can be represented as follow:

$$Fe_3O_4(s) + H_2(g) = 3FeO(s) + H_2O(g) \quad \Delta H = 63.7 \text{ kJ}; K_{eq} = 0.0117 \text{ at } 673 \text{ K} \tag{1}$$

$$FeO(s) + H_2(g) = Fe(s) + H_2O(g)\,) \quad \Delta H = 16.6 \text{ kJ}; K_{eq} = 0.147 \text{ at } 673 \text{ K} \tag{2}$$

For simplicity, FeO represents the nonstoichiometric nature of wüstite ($Fe_{1-x}O$). The milled concentrate was tested to find the optimal experimental parameters that lead to the chemical reaction as the slowest mechanism. Thus, to observe the effect of diffusion through the particles, the weight of the sample was varied as shown in Figure 5. The fractional reduction, X, was defined as the ratio of the instant weight loss, ΔWt, over the theoretical final weight loss, $\Delta W\infty$, corresponding to the loss of four oxygen atoms per Fe_3O_4 unit. It can be seen that from about 4 mg and below the reaction rate seems to remain constant. It should be remarked that 2 mg of powder did not cover the whole surface area of the crucible; therefore, 2 mg of powder was considered to study the reduction kinetics of magnetite concentrate by hydrogen.

Similarly, to examine the aspect of external mass transport in the gas phase, different flow rates of hydrogen were tested as shown in Figure 6. A hydrogen flow rate of 120 mL/min would avoid mass transfer issues on the reduction process.

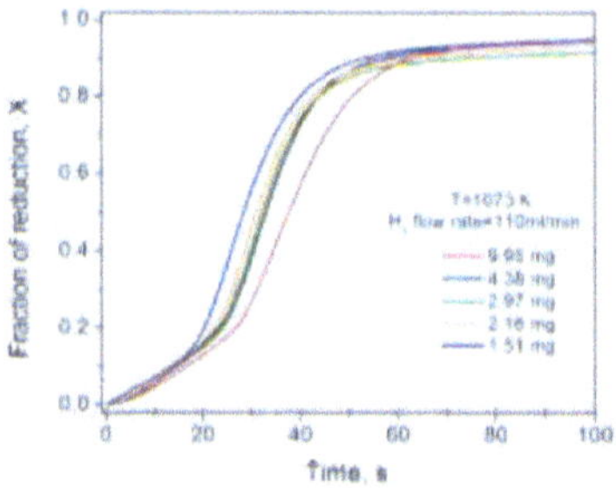

Figure 5. Effect of sample weight on the reduction reaction of 8-h milled magnetite concentrate.

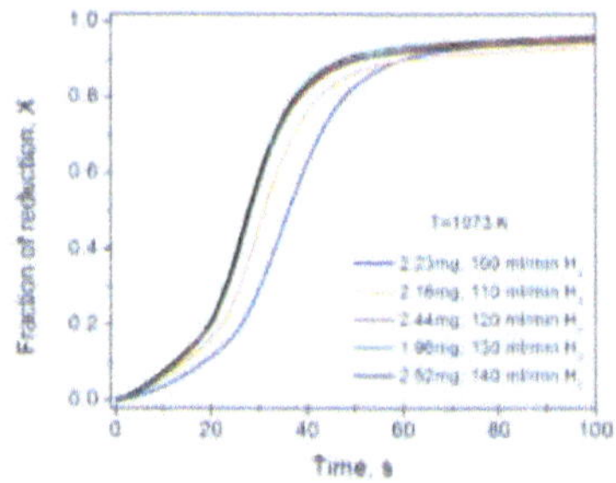

Figure 6. Effect of hydrogen flow rate on the reduction reaction of 8-h milled magnetite concentrate.

On the other hand, since the mean particle size for the 8-h-milled concentrate is about 4 μm and a very thin layer of loose particles was used, it is expected that the heat transfer would not be the controlling step even though the endothermic heats of Reactions (1) and (2) are appreciable.

From the above results, it is reasonable to assume that using about 2 mg of powder with a hydrogen flow rate of 120 mL/min, the rate-controlling step for Reactions (1) and (2) is most likely to be the chemical reaction at the reaction interface, except perhaps at the final stage of reduction.

Figure 7 shows the fractional reduction rate for the isothermal reduction of milled magnetite concentrate by hydrogen in the temperature range of 623 K to 1123 K. There is a detectable induction period at the beginning of the reaction which may indicate an intermediate reaction as aforementioned. As expected, the reduction rate increases with the increase in temperature being more evident in the temperature range of 673 K to 823 K. It is clearly seen that, at 623 K, the reduction rate is extremely slow. In fact, after 1800 s of processing the reduced fraction was estimated to be only 0.6. While at 1123 K, it only takes 45 s to reach a reduce fraction of 0.86. This low rate may indicate sintering of particles leading to a decrease in surface area.

It can also be noticed that at 973 K, the reduction rate is slightly slower than that at 873 K. This observation can be attributed to significant sintering of particles.

The isothermal reduction experiments of as-received concentrate are shown in Figure 8. It is observed that only the experiments conducted at 673 K, 723 K, and 773 K are more sensitive to temperature increase. This behavior is not uncommon distinct in chemical-reaction controlled processes. At higher temperatures, the increase in the reduction rate with temperature drops dramatically. Generally, all reduction curves in Figure 8 present a slower reduction kinetics compared with the milled concentrates, which is expected due to the larger particle size of the as-received ore.

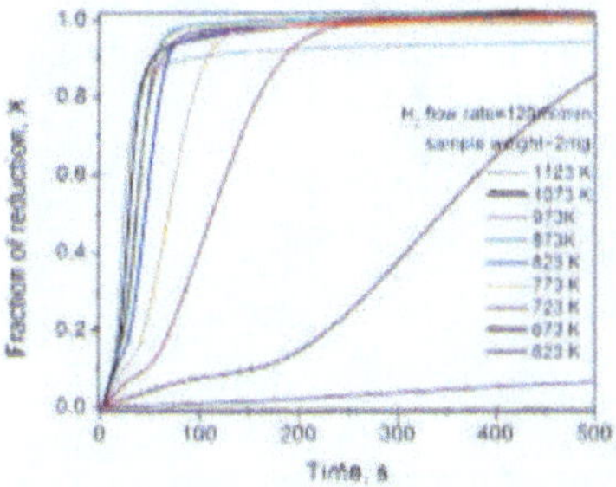

Figure 7. Effect of temperature on the reduction reaction of the 8-h milled magnetite concentrate.

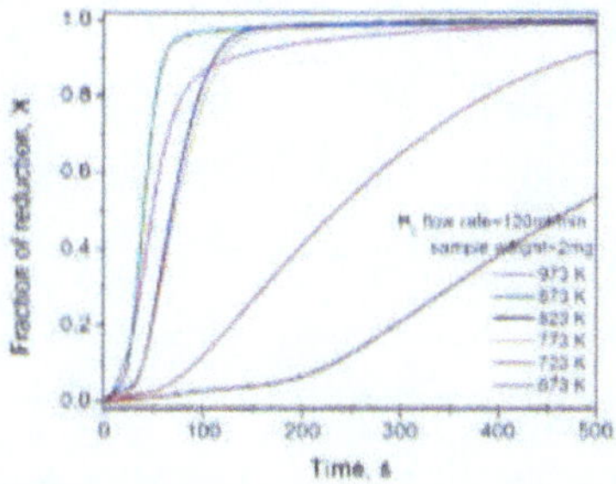

Figure 8. Effect of temperature on the reduction reaction of the as-received magnetite concentrate.

Figure 9 shows SEM images of milled particles and as-received particles before and after being reduced at 1123 K. Before the reduction process the as-received particles show an irregular and rectangular shape while the milled particles have a nearly spherical shape with tendency to form agglomerates. The difference in particle size between the two samples is very evident. The SEM images of both samples after being reduced at 1123 K reveal structures with less homogeneous porosity which can be attributed to sintering. The sintering effect was already elucidated in the reduction curves at higher temperatures as discussed earlier.

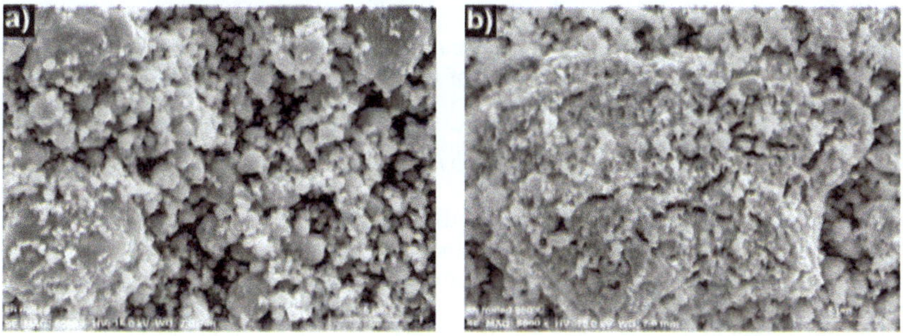

Figure 9. SEM images of magnetite concentrate a) milled for 8 h, b) milled for 8 h after being reduced at 1123 K, c) as-received, and d) as-received after being reduced at 1123 K.

The nucleation and grow kinetics has been found to describe the rates of reduction of iron oxide [6] and cuprous sulfide [7]. Recently, Wang and Sohn [4] used the nucleation and growth model to develop a rate expression for the hydrogen reduction of magnetite concentrates, in the temperature range of 1423 - 1673 K, by means of a novel gas-solid suspension set up. The nucleation and grow expression attributed to Avrami [8] is represented as:

$$[-\mathrm{Ln}(1-X)]^{1/m} = k_{app}\, t \qquad (3)$$

where X is the fraction of reduction, t is the reaction time, m is a constant and k_{app} is the apparent rate constant which contains the dependence of the rate on gaseous reactant concentration and thus can be written as [9]:

$$k_{app} = b\, k f(C_{H2}) \qquad (4)$$

where b is the number of moles of solid reacted by one mole of gaseous reactant (b = 1 in this work), k is the intrinsic rate constant, C_{H2} is the hydrogen concentration in the bulk phase in mol/cm^3, and f designates the rate dependence on C_{H2}. All experiments were conducted at very excess driving force therefore the reverse reaction was not considered. Assuming a first-order reaction with respect to hydrogen concentration, namely $f(C_{H2}) = C_{H2}$, Equation (4) yields to $k = k_{app}/C_{H2}$. From Equation (3), It is seen that Ln[-Ln(1-X)] should be a linear function of Ln t with m as the slope and m Ln k_{app} as the intercept with the Ln t = 0 axis.

Figure 10 shows a plot of Ln[-Ln(1-X)] against Ln t for the results shown in Figure 8 with the isothermal data corresponding to $0.2 < X < 0.9$. The average value of the slopes obtained by regression analysis was 2.89 confirming that nucleation and grow products were three-dimensional. To obtain the intercept with the Ln t = 0 axis, and thus the value of k_{app}, a constant value of m = 3 was used throughout. Figure 11 a plot of Ln[-Ln(1-X)] against Ln t for the results shown in Figure 9. In this case, the average value of the slopes was 2.61 which is in accordance with the less spherical particle shape observed in SEM images. For simplicity, a value of $m = 3$ was used to obtain the apparent rate constants.

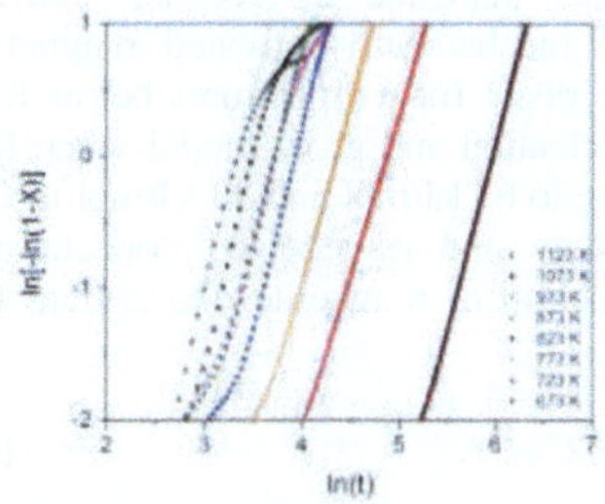

Figure 10. Plot of Ln[-ln(1-X)] *vs* Ln t from the results of Figure 8.

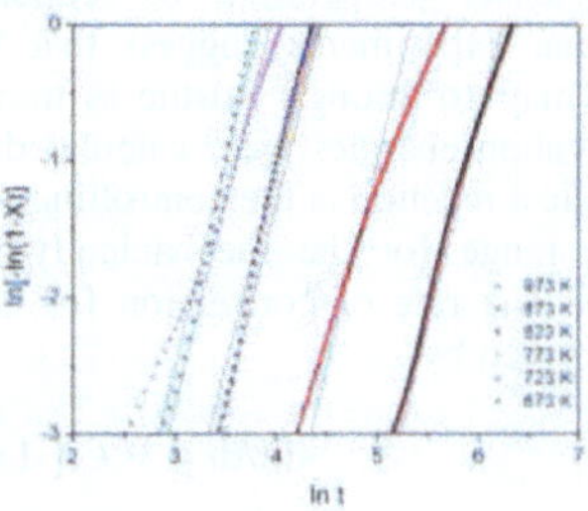

Figure 11. Plot of Ln[-Ln(1-X)] *vs* Ln t from the results of Figure 9.

Figure 12 shows the intrinsic rate constants as an Arrhenius plot. The experimental data seem to fit different consecutive slopes. Similar mode has been also reported in Ref. [1]. The rate constants, corresponding to higher temperatures, were left out from the calculation of the activation energy as it is likely to follow a different reduction mechanism. The computed values of the activation energy led to 67 kJ/mol and 80 kJ/mol for the mechanical activated concentrate and as-received concentrate, respectively. The activation energies thus obtained would correspond to the chemical reaction as the rate controlling as only those rate constants that were sensitive to temperature increase were considered. The difference in the activation energies should be attributed to the mechanical activation alone and not to the difference in particle size. A change in particle size can modify the rate constant at different temperatures but not the value of activation energy provided the system is controlled by the rate of the chemical reaction. Although the difference in activation energy does not appear to be significant, it should be born in mind that the equilibrium constant for Reaction (2) is less than one; therefore, the activation energy tends to be small and so is the difference in activation energies.

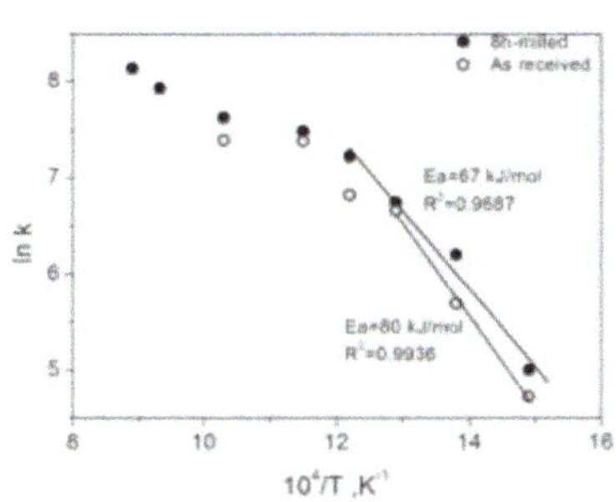

Figure 12. Arrhenius plots for the isothermal reduction of 8-h-milled and as-received magnetite concentrate.

Conclusions

In the present work, the mechanically activated concentrate with a lattice micro strain 0.30 showed an onset temperature of reduction 100 K lower than the as-received sample. Nonisothermal experiments suggest that the reduction of mechanically-activated magnetite concentrate may go through wüstite as transitory metastable phase for temperatures below 853 K. The activation energies were calculated based on the nucleation and grow model when the rate of chemical reaction is the controlling mechanism leading to 67 kJ/mol and 80 kJ/mol in the temperature range for the mechanically-activated concentrate and as–received concentrate, respectively. The rate of conversion for the hydrogen reduction of magnetite concentrate by hydrogen is given by:

$$dX/dt = 3 \cdot k \cdot [-\mathrm{Ln}(1-X)]^{2/3} \cdot (1-X) \cdot C_{H2} \quad (5)$$

with $k = 2.96\times10^{6}\cdot\exp(-8058/T)$ ($cm^3\cdot mol^{-1}\cdot s^{-1}$) for the 8-h-milled magnetite concentrate, and $k = 1.94\times10^{8}\cdot\exp(-9622/T)$ ($cm^3\cdot mol^{-1}\cdot s^{-1}$) for the as-received magnetite concentrate.

References

1. A. Pineau, N. Kanari and I. Gaballah, "Kinetics of reduction of iron oxides by H_2 Part II. Low temperature reduction of magnetite," Thermochimica Acta, 456 (2007), 75–88.
2. P. Pourghahramani and E. Forssberg, "Effects of mechanical activation on the reduction behavior of hematite concentrate," Int. J. Miner. Process, 82 (2007), 96–105.
3. A. Pineau, N. Kanari and I. Gaballah, "Kinetics of reduction of iron oxides by H_2 Part I. Low temperature reduction of hematite," Thermochimica Acta, 447 (2006), 89–100.
4. H. Wang and H.Y. Sohn, "Hydrogen Reduction Kinetics of Magnetite Concentrate Particles Relevant to a Novel Flash Ironmaking Process," Metall. Mater. Trans. B, 44.1 (2013), 133-145.
5. R. Morales, Du Sichen and S. Seetharaman, "Reduction Kinetics of Fe_2MoO_4 Fine Powder by Hydrogen in a Fluidized Bed," Metall. Mater. Trans. B, 34.10 (2003), 661-667.
6. Y. K. Rao, "Mechanism and the intrinsic rates of reduction of metallic oxides," Metall. Mater. Trans. B, 10 (1979), 243-255.
7. H. Y. Sohn and S. Won, "Intrinsic kinetics of the hydrogen reduction of Cu_2S," Metall. Mater. Trans. B, 16 (1985), 831-839.
8. M. Avrami, "Granulation, Phase Change, and Microstructure Kinetics of Phase Change. III," J. Chem. Phys., 9 (1941), 177-184.
9. H. Y. Sohn, "The law of additive reaction times in fluid-solid reactions," Metall. Mater. Trans. B, 9 (1978), 89-96.

EPD Congress 2016
Edited by: Antoine Allanore, Laura Bartlett, Cong Wang, Lifeng Zhang, and Jonghyun Lee
TMS (The Minerals, Metals & Materials Society), 2016

INFLUENCES OF THERMOMECHANICAL PROCESSING ON THE MICROSTRUCTURE AND MECHANICAL PROPERTIES OF A HSLA STEEL

Yu Zhao, Songsong Xu, Yun Zou, Jinhui Li, Z.W. Zhang*

Key Laboratory of Superlight Materials and Surface Technology, Ministry of Education, College of Materials Science and Chemical Engineering, Harbin Engineering University, Harbin 150001, P R China

Keywords: HSLA steels, Microstructure, Mechanical properties, Thermomechanical processing, Nano-scale precipitation

Abstract

High strength low alloy (HSLA) steels with high strength, high toughness, good corrosion resistance and weldability, can be widely used in shipbuilding, automobile, construction, bridging industry, etc. The microstructure evolution and mechanical properties can be influenced by thermomechanical processing. In this study, themomechanical processing is optimized to control the matrix microstructure and nano-scale precipitates in the matrix simultaneously. It is found that the low-temperature toughness and ductility of the steels are significantly the matrix microstructure during enhancing the strength by introducing the nano-scale precipitates. The effects of alloying elements on the microstructure evolution and nano-scale precipitation are also discussed.

Introduction

High strength low alloy (HSLA) steel is developed on the basis of the research on ordinary carbon steel. Because of the high copper content in the steel, strength can be greatly improved by the precipitation of nano-phases. At the same time, the plastic property of the material can also be guaranteed. Meanwhile, the price of the steel is reasonable. So these kinds of steels can be widely used in the automobile industry, shipbuilding industry, marine engineering and so on [1-2]. Recently, HSLA steels with excellent properties has been required greatly, especially in the aspects of welding performance and low temperature toughness. HSLA steel with copper precipitation strengthening has received a wide range of attention, due to promising applications.

In 1930s, researchers have found that the strength of steel could be improved by adding copper [3]. When heat treatment temperature was set in the range of 450℃-600℃, it could lead to the precipitation of Cu element, and the yield strength could be increased by 100-200MPa. In 1980s, the HSLA-80 steel first appeared [4], and then was updated to HSLA-100 and HSLA-115 later. The yield strength of HSLA-80 steel is the same as that of HY-80 steel, about 550 MPa [5, 6]. It is shows that when the size of the nano-phase is 2.5-3.0nm, the precipitation hardening peak could be obtained [7, 8]. In general, when the nano particles grow to a larger size is larger size (>10nm), the crystal structure of the particles transforms form the body-centered cubic (BBC) structure to the face centered cubic (FCC) structure [8-10]. Some scholars also found that by optimizing the content of alloy elements and themomechanical processing, a proper copper rich precipitates and matrix microstructure can be obtained [11], leading to excellent mechanical properties [12,13].

During the preparation of HSLA steel, controlling both the matrix microstructure and copper rich precipitates becomes the main concerns of the process. The aim of this study is to elucidate the effect of aging temperature and time on the mechanical properties and microstructure of the HSLA steel.

Experimental

To ensure the accuracy of the experiment, and avoid the contamination of the impurity of the material, high purity alloying elements were used(Fe, 99.9%; Mn, 99.9%; Ni, 99.9%; Cu, 99.99%; Al, 99.99%; Ti, 99.995%; B-Fe(boron content, 17%;). The alloy composition is shown in Table 1

Table 1 Chemical compositions of steel (wt.%)

Fe	Cu	Mn	Ni	Al	B	Ti
bal	2.0%	1.0%	4.0%	1.0%	0.01%	0.30%

The cast ingots were obtained using high vacuum arc melting under the protection of high purity argon (>99.95%).The ingots were remelted several times with the magnetic stirring to ensure homogeneous distribution of alloying elements (180mm×20mm i.d.). The as-cast specimens with a diameter of 20 mm and 180 mm long were heated at 1000 ℃ in air for 1h, and then hot rolled to a total thickness reduction of ~80%, followed by water quenching. The hot rolled samples were then cold to a thickness reduction of ~50% after removing the surface oxides. The final size of as-rolled samples are 2mm. The as-rolled samples were solid solution treated at 900℃ for 30min, then quenched in water. Then the materials were aged at 500℃, 550℃ and 600 ℃ for 1h, 2h, 5h and 10h, respectively. The average Vickers hardness measurement was conducted applying 1000g load, and the samples were test over 10 locations. The sheet tensile samples were processed by electro-discharge machining (EDM) with a gauge size of 20.26mm×2mm×5mm. The tensile tests were conducted on INSTRON 5565 at room temperature with a strain rate of $\sim 1\times 10^{-3}$/s. The phase components of the specimens were determined by X-ray diffraction (XRD) with the following experimental parameters: a Cu Target K alpha ray, tube pressure is 40kV tube pressure, 20°-90°scanning angle, and 5°/min of scanning speed.

Optical microscopy (OM) and scanning electron microscope (SEM) were also used to characterize the microstructure of the specimens. The fracture surfaces of tensile samples were examined by SEM to evaluate the type of fracture.

Results and Discussion

Microstructure observation

The microstructures of the as-rolled and solid-solution-treated samples are show in Fig.1 After rolling, fibrous tissue is formed with fiber direction paralleling to the rolling direction (Fig.1a). After heat treatment at 900℃for 1h, the deformed fibrous structure transforms into irregular polygon ferrite due to recrystallization (Fig.2b).

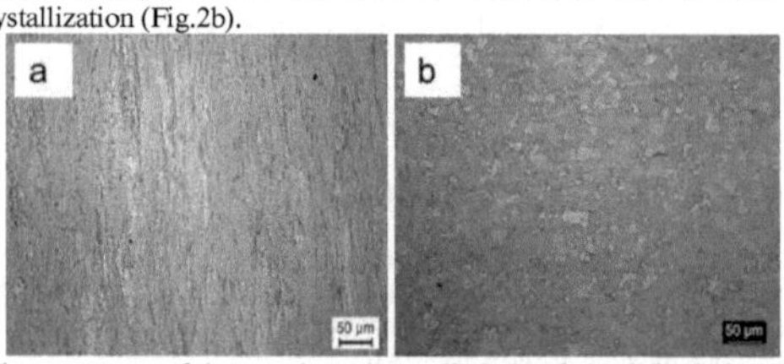

Fig.1 Microstructures of the samples, (a) as-rolled, (b) after solid solution treatment

Figure 2 shows the microstructures of the samples aged at 500℃ for 1h, 2h, 5h and 10h, respectively. It can be seen that the samples are mainly composed of polygonal ferrite with a average grain size.

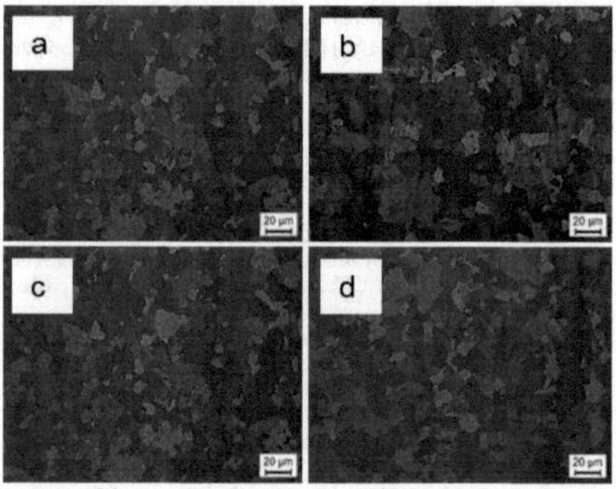

Fig.2 Micrstructures of the samples after aging at 500℃ for (a) 1h, (b) 2h, (c)5h and (d)10h

Figure 3 and 4 shows the microstructures of the samples aged at 550℃ and 600℃ for 1h, 2h, 5h and 10h, respectively. After aging at 550℃ and 600℃, the microstructures are similar to that after aging at 500℃, while the grain size grows slightly to ~20 μm and ~30 μm for 550℃ and 600℃, respectively. It can be concluded from Fig.2-4 that aging at different temperatures do not change the microstructures significantly.

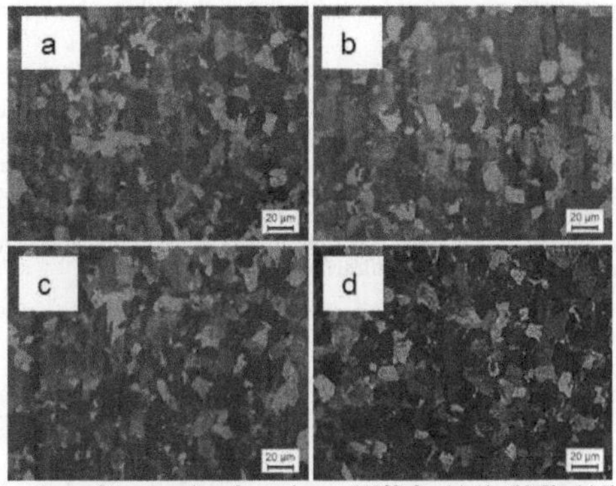

Fig.3 Micrstructures of the samples after aging at 550℃ for (a) 1h, (b) 2h, (c)5h and (d)10h

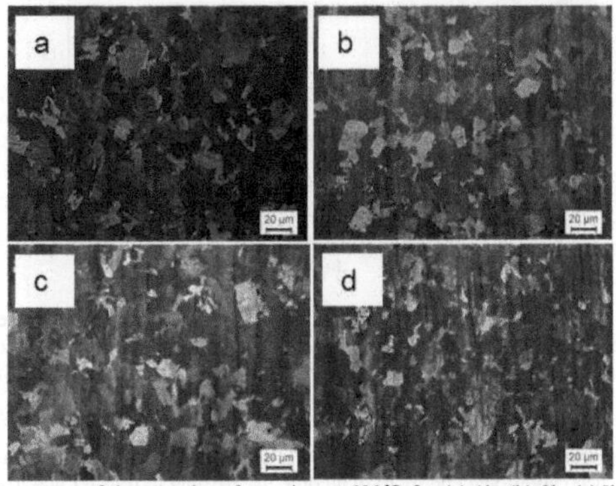

Fig.4 Micrstructures of the samples after aging at 600℃ for (a) 1h, (b) 2h, (c)5h and (d)10h

Phase analysis

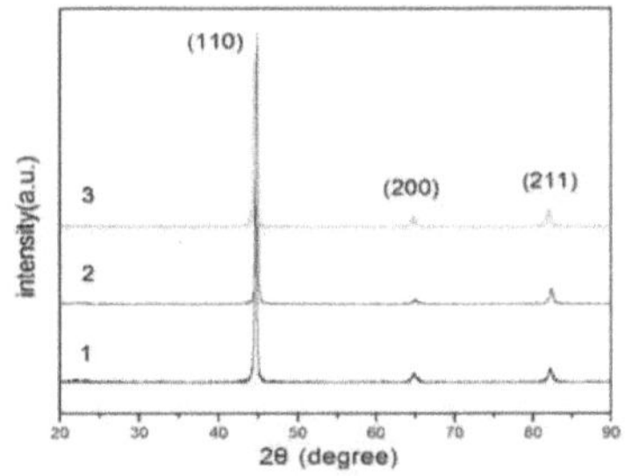

Fig.5 X ray diffraction spectra for different states of HSLA steel
(1-as-rolled; 2-solid solution treated; 3 after aging)

Figure 5 shows the diffraction spectra of HSLA steel after various treatments. It can be seen that the samples displays similar diffraction spectra, where only three peaks representing ferrite are found, revealing that no phase change has happened during various processing.

Mechanical properties

The microhardness of samples after various treatments are shown in Fig.6. After solid-solution-treated, the microhardness of the samples is ~267 HV. Upon aging, the microhardness increases significantly. With the three aging temperatures, three aging stage can be observed, i.e. under aging, peak aging and over aging. After aging at 500℃, the samples have the highest microhardness and the aging peak appears at 5 h. Increasing the aging temperature, the microhardness drops and the aging peak moves towards to short aging time. These phenomena are similar to the traditional aging processes in aged-strengthened alloys. The changes in microhardness can be attribute to the formation of precipitates in the steels. After solid-solution treating, there is few precipitates in the matrix, leading to a low microhardness. Upon aging, the nucleation and growth of the precipitates leads to the significant increase in hardness.

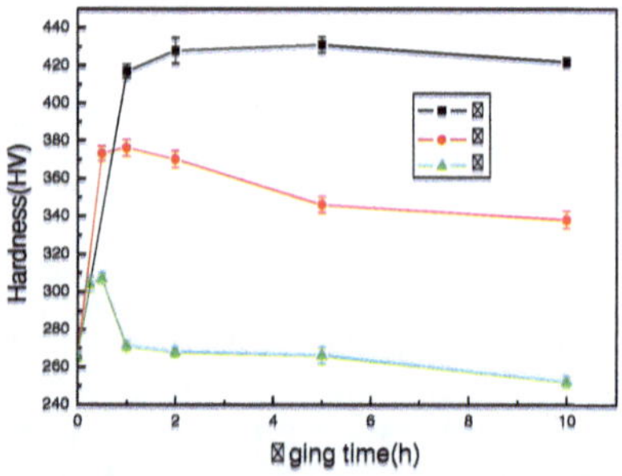

Fig.6 The microhardness of samples (the time 0h is solid solution sate) aged at different temperature(500℃, 550℃, 600℃) for 1h, 2h, 5h and 10h

When aging temperature was set at 550℃ and 600℃, the aging peak appeared at 1 h and 0.5h, respectively (Fig.6). With the increase of aging temperature, the aging peak of HSLA steel moves towards shorter aging time, and the hardness of samples decreases as well. This is because aging temperature directly affects the diffusion coefficient of copper in the matrix, following the diffusion formula:

$$D = D_0 \exp(-\frac{Q}{RT}) \tag{1}$$

Where R is the gas constant, T is temperature, D is the diffusion coefficient, Q indicates the diffusion activation energy. When the copper element diffuses in the crystal, they need additional energy to overcome the barriers to transfer and to strengthen the material [16]. It can be seen from Eq. (1) that the diffusion coefficient D and temperature T is exponential interrelated. When increasing the aging temperature, the copper atoms will diffuse quickly and so does the precipitation rate. Then the aging peak anticipated.

The tensile properties under various aging temperature and time are show in Fig.7. Fig7-a describes the tensile properties of samples aged at 500℃ for 1h, 5h, 10h and solid solution. From Figure 7-a, it can be seen that the highest tensile properties of 1206MPa can be obtained after aging for 5 h. Zhang Z W [17] has reported that aging process leads to the formation of nano-scale Cu-rich precipitates, which can enhance the strength significantly by pinning the dislocation during deformation.

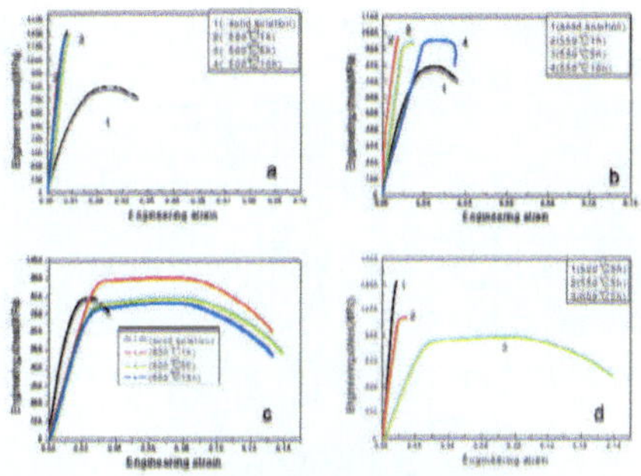

Fig.7 Tensile properties of steel at various aging temperature and time

From figure 7-b, it can be concluded that the strength of sample reaches the peak at 550℃for 1h (983MPa). And as aging time prolonged, the strength of the steels decreased with an increase in elongation. Figure 7-c shows the tensile stress-strain curves of samples after aging at 600℃. It shows that the plastic properties of the samples after different aging time were all significantly improved. The extension rate increases to 14% after aging for 5 h. When aging for 10 h, both the elongation and strength decreased. This is because the higher temperature and longer time lead to the rapid growth of nano-sized precipitates [18]. Taking overall mechanical a property into account, the better steel is the one aged for 1h, whose strength is 900MPa and percentage elongation is 13%.

Fig.7-d shows the effect of aging temperature on the tensile properties. As the aging temperature increases, the strength of the material decreases, and the plasticity of the material gradually increases. This is because the growth rate of the precipitates accelerated and its ability to block the dislocation movement is reduced, resulting in a decrease of the hardness and strength. The deformation mode of the material transferred from brittle deformation to plastic deformation.

Fracture analysis

Figure 8 shows the fracture surface of the sample after solid-solution treatment. It can be seen that the fracture morphology of the HSLA steel at solid solution state is a dimple pattern. As we know, the mainly micro-morphology of the ductile fracture metal is the existence of the dimples which are formed by the aggregation of the holes. When the external load is applied, these cavities went through nucleating, growing up and then gather to the fracture. It is clear that the fracture mode of the solid solution is microvoid coalescence fracture, which belongs to plastic fracture.

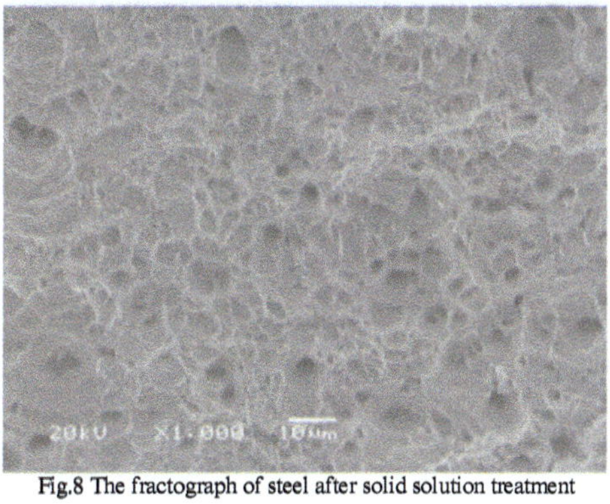

Fig.8 The fractograph of steel after solid solution treatment

Figure. 9 shows the fracture surface of samples aged at 500 ℃ with various aging time. It can be seen from Fig.9 that all the samples show a brittle intergranular fracture along with a few of cleavage fracture.

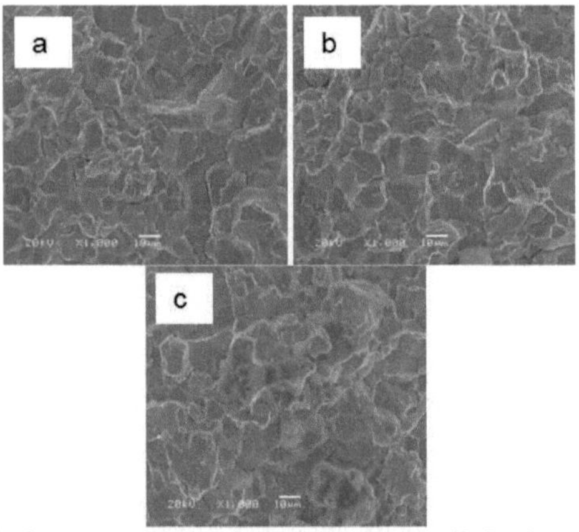

Fig.9 Tensile fracture morphologies of the samples aged at 500 ℃ for 1h (a), 2h (b) and 5h (c)

As shown in Fig.10, the fracture surface of the steel is still typical intergranular fracture after aging at 550℃ for 1h and 5h (fig. 10a and b). When aging for 10h, the fractograph exhibits a small part of cleavage step and a few dimples, contributing some plasticity.

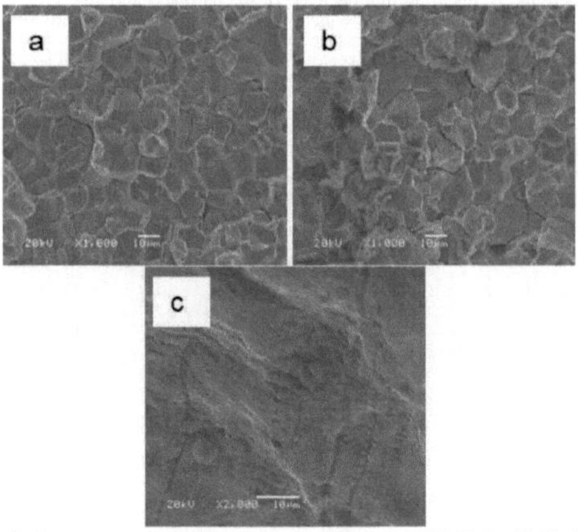

Fig.10 Tensile fracture morphologies of the sample aged at 550 ℃ for 1h (a), 2h (b) and 5h (c)

When aging process is set at 600℃(Fig. 11 a-c), the tensile fracture mode of the material is typical microvoid coalescence fracture with different size of dimples. Therefore, the elongation of the material is improved greatly.

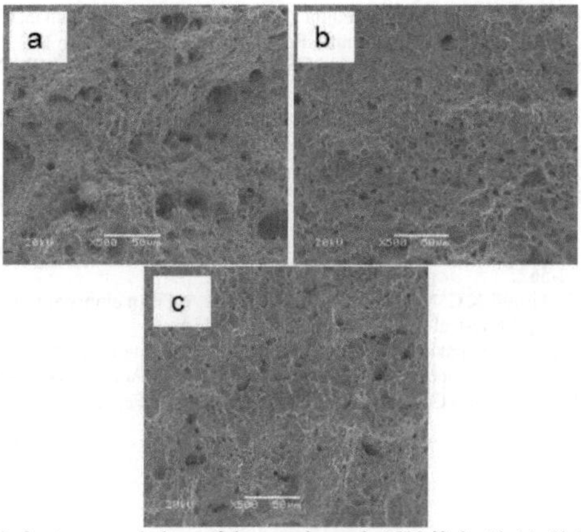

Fig. 11 Tensile fracture morphology of the sample aged at 600 ℃ for 1h (a), 2h (b) and 5h (c)

Conclusions

1 After rolling treatment, the microstructure is elongated and aligned along the rolling direction. After the process of solution treatment at 900℃ for 1h, the microstructure is polygonal ferrite. Aging treatment has no significant effect on the microstructure.

2 Aging treatment can improve the hardness of the metal significantly. With increasing the time of aging, the aging peak appears earlier, and the hardness decreases.

3. At the same temperature, the change in the strength under various aging time is consistent to the change in hardness. At 500℃ for 5 h, the maximum strength was obtained (1206MPa). At 550℃ and 600℃, the maximum strength appeared at 1h, which were 983MPa and 900MPa, respectively. With the increase of aging time, the elongation rate also increased. When aging time is the same, the tensile strength decreased and the elongation increased with the increase of aging temperature.

4. When aging temperature is 500℃, increasing aging time does not change the fracture mode. All of them are intergranular fracture. When aging at 550℃ for 1h and 2h, the fracture mode of the material is also intergranular fracture. 10 h of aging at 550℃ leads to the cleavage fracture. When the aging time goes up to 600 ℃, the fracture mode of the material transfers to ductile mode, i.e. microporous gathered fracture mode.

Acknowledgements

This research was supported by the Fundamental Research Funds for the Central Universities (HEUCFZ1308), the NSFC Funding (51171081, 51371062 and U1460102), NSFHLJ (ZD201411), the Scientific Research Foundation for the Returned Overseas Chinese Scholars, Heilongjiang province and the Project 2013DB04 at NPL, CAEP. Yu Zhao is benefited from the International Exchange Program of Harbin Engineering University for Innovation-oriented Talents Cultivation.

References

[1]Xu Zu yao Self-reliance innovation on development of ultra-high strength steel [J]. Shanghai Metals, 2009, 31(2): 1-6.

[2] Yong GAN and Han DONG. Recent Progress in Advanced Steel Technologies [J], China Metallurgy, 2004(8): 1-6.

[3] Kinnear H B. One percent copper steel has many desirable physical qualities [J]. Iron Age 1931, 10: 696-672.

[4] Montemarano T W, Sack B P, Gudas J P, Vassilaros M G, Vanderveldt H H. High strength low alloy steels in naval construction[J].Journal of Ship Production,1986,3:145-152.

[5] Fine M E, Liu J Z, Asta M. D. An unsolved mystery: The composition of bcc Cu alloy precipitates in bcc Fe and steels [J]. Materials Science and Engineering A, 2007, A463:271–274.

[6] Goodman S R, Brenner S S and Low J R. An FIM-atom probe study of the precipitation of copper from lron-1.4 at. pct copper. Part II: Atom probe analyses [J].Metallurgical Transactions, 1973, 4(10): 2371-2378.

[7] Goodman S R, Brenner S S and Low J R. An FIM-atom probe study of the precipitation of copper from lron-1.4 at. pct copper. Part I: Field-ion microscopy [J].Metallurgical Transactions, 1973, 4(10): 2363-2369.

[8] Othen P J, Jenkins M L, Smith G W D. High-resolution electron-microscopy studies of the structure of Cu precipitates in alpha-Fe[J], Philosophy Magazine A,1994, A70(1):1-24.

[9] Othen P J, Jenkins M L, Smith G D W. Transmission electron microscope investigations of the structure of copper precipitates in thermally-aged Fe—Cu and Fe—Cu—Ni[J]. Philosophical Magazine Letters, 1991, 64(6):383-391.

[10] Zhang Z W, Liu C T, Miller M K, Wang X -L,, Wen Y R, Fujita T, Hirata A, Chen M W, Chen G, Chin B A.A nanoscale co-precipitation approach for property enhancement of Fe-base alloys[J], Scientific Reports, 2013, 3: 1327.
[11] Lu K, Lu L, Suresh S. Strengthening Materials by Engineering Coherent Internal Boundaries at the Nanoscale [J]. Science, 2009, 324(17): 349-352.
[12] Fine M E, Liu J Z, Asta M D. An Unsolved Mystery: the Com-position of bcc Cu Alloy Precipitates in bcc Fe and Steels [J]. Materials Science and Engineering: A, 2007, 463: 271-274.
[13] Liu C T. Recent Advances in Ordered Intermetallics [J] Materials Chemistry and Physics, 1995, 42: 77-86.
[14] Teng Z K, Liu C T, Ghosh Getal. Effects of Al on the Micro-structure and Ductility of NiAl-Strengthened Ferritic Steels at Room Temperature [J]. Intermetallics, 2010, 18: 1 437-1 441.
[15] Schobera M, Schnitzerb R, Leitner H. Precipitation Evolutionin the Ti-Free and Ti-Containing Stainless Maraging Steel [J]. Ultramicroscopy, 2009, 109: 553-562.
[16] Zhang Z W, Liu C T, Wang X L, Littrell K C, Miller M K, An K, Chin B A. From embryos to precipitates: A study of nucleation and growth in a multicomponent ferriticsteel [J].Physical Review B, 2011, 84: 174114.
[17] Kapoor Monica, Isheim D, Ghosh G, Vayanman S, Fine M E, Chung Y W. Aging characteristics and mechanical properties of 1600 MPa body-centered cubic Cu and B2-NiAl precipitation-strengthened ferritic steel[J]. Acta Materialia, 2014, 73: 56-74.
[18] Yen H W, Huang C Y, Yang J R. The Nano Carbide Control: Design of Super Ferrite in Steels [J]. Advanced Materials Research, 2010, 89-91: 663-668.

EPD Congress 2016
Edited by: Antoine Allanore, Laura Bartlett, Cong Wang, Lifeng Zhang, and Jonghyun Lee
TMS (The Minerals, Metals & Materials Society), 2016

BEHAVIORS AND EVOLUTIONS OF $MgO•Al_2O_3$ IN NON-ORIENTED SILICON STEEL DURING CALCIUM TREATMENT

Yong Zhao, Yan-hui Sun
Collaborative Innovation Center of Steel Technology, University of Science and Technology Beijing 100083, China

Keywords: $MgO•Al_2O_3$; calcium treatment; non-oriented silicon steel; behavior; evolution

Abstract

Behaviors and evolutions of $MgO·Al_2O_3$ in non-oriented silicon steel during calcium treatment were investigated by thermodynamic calculations and high-temperature experiments. $MgO·Al_2O_3$ inclusions were the main type of inclusions before calcium treatment, which is also confirmed by the stability diagram of $MgO/MgO·Al_2O_3/Al_2O_3$ from thermodynamic calculations. After calcium treatment, the majority of original $MgO·Al_2O_3$ inclusions were mainly modified to CaO-MgO-Al_2O_3 system inclusions, this would be beneficial to improve castability and product quality of silicon steel. But there was no modification happening to some $MgO·Al_2O_3$ spinels which were directly surrounded by CaS inclusions. The content of Al_2O_3 occasionally existed a sudden increase besides a common decrease in the product layer of $CaO·MgO·Al_2O_3$. Then the evolution mechanisms of $MgO·Al_2O_3$ inclusions were proposed in present work. Thus, Al/Mg atom ratio had a significant influence on the modification of $MgO·Al_2O_3$ spinel.

Introduction

Non-oriented silicon steels are mainly applied in engines or motors. The cleanliness and product quality of steel are restricted by the existence of nonmetallic inclusions, such as nitrides, oxides, and sulfides, which are detrimental to the continuous casting of steel and easy to cause some kinds of weakness [1-2]. In order to control the characteristics of inclusions, such as type, shape, size, and composition, calcium treatment is widely used to transform high melting temperature inclusions into low melting temperature inclusions.

$MgO·Al_2O_3$ spinel was the important type of inclusions in molten steel, many researchers have studied the formation and modification mechanism of $MgO·Al_2O_3$ inclusions. Park *et al.* [3-4] proposed the ionic reaction model in inclusion modification process and consider the "MgO to spinel transformation" mechanism using unreacted model including ionic diffusion mechanism. Deng *et al.* [5] proposed that inclusions changed with the route as $Al_2O_3 \rightarrow MgO·Al_2O_3 \rightarrow CaO·MgO·Al_2O_3$ which was finally surrounded by a layer of $CaO·Al_2O_3$ according to industrial experiments and thermodynamic calculations. Guo *et al.* [6] used the calculation software FactSage to analyze the formation of Al_2O_3, $MgO·Al_2O_3$ spinel, and various calcium aluminates which were also studied by industrial experiments. Pretorius *et al.* [7] and VERMA *et al.* [8] proposed that small but nonzero concentrations of Magnesium in calcium modified inclusions could contribute significantly to liquefaction of the inclusions and promote castability. Of course, ITOH *et al.* [9], FUJII *et al.* [10], Yang *et al.* [11], and Kang *et al.* [12] also studied the formation and modification of $MgO·Al_2O_3$ inclusions by using thermodynamic calculations, laboratory experiments or industrial tests. Laboratory experiments regarding calcium treatment of non-oriented silicon steel were carried out in the present work, and evolution mechanisms of $MgO·Al_2O_3$ inclusions were investigated, as will be detailedly described in the following article.

Experimental methods

In the present work, a 50-kW and 10-kg vacuum induction furnace with a magnesia crucible was used to melt the whole raw materials which include pure iron, electrolytic manganese, ferrosilicon, aluminium particles, calcium silicon wire, refining slag (50 mass% CaO, 5 mass%

SiO_2, 30 mass% Al_2O_3, 10 mass% MgO, 5 mass% CaF_2). The pure iron of about 6 kilograms and the refining slag of about 300 grams were charged into the crucible.

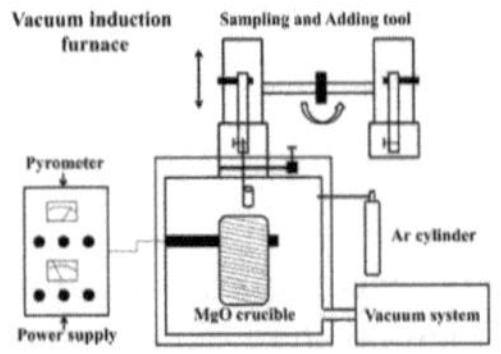

Figure 1. Experimental setup

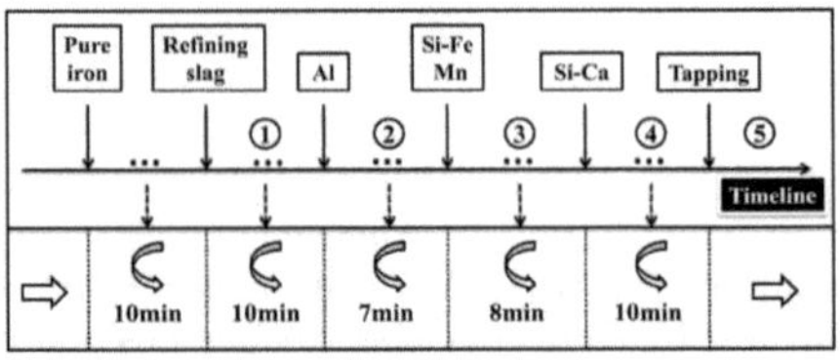

Figure 2. Experimental procedure diagram

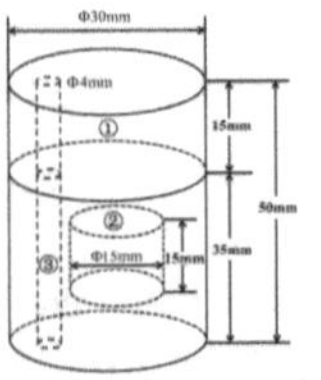

Figure 3. Sample processing

The Schematic of experimental setup and sampling tool was shown in Figure 1. The melt temperature was controlled to be 1873K using a thermocouple. The experimental procedure diagram was shown in Figure 2. During the experiment, steel samples were taken before and after calcium treatment (①-⑤ in Figure 2), and quickly quenched in water. Steel samples (Φ30mm×50mm) extracted from the molten steel were processed for three parts (shown in Figure 3). Part 1(Φ30mm×15mm, ① in Figure 3) was used to analyze the chemical compositions of the bulk by ICP-AES method, part 2 (Φ15mm×15mm, ② in Figure 3) and part 3 (Φ4mm×50mm, ③ in Figure 3) were prepared to obtain information of inclusions by using SEM-EDS and oxygen content by inert gas pulse infrared thermal method, respectively.

Results and discussion

Chemical Compositions of Molten Steel and Inclusions

Chemical compositions of the molten steel were given in Table I. It showed the mass percent of the main elements, such as carbon, silicon, manganese, aluminium, calcium, magnesium and sulfur. The content of the total oxygen and nitrogen were also given in Table I.

Table I. Chemical Compositions of Molten Steel, mass%

Number	C	Si	Mn	Al	Ca	S	Mg	T[O]	N
①	0.005	0.021	0.024	0.015	0.0004	0.003	0.0006	0.0520	0.0076
②	0.006	0.023	0.027	0.251	0.0003	0.003	0.0005	0.0025	0.0078
③	0.006	2.462	0.294	0.248	0.0005	0.004	0.0005	0.0023	0.0079
④	0.006	2.459	0.289	0.248	0.0015	0.003	0.0006	0.0023	0.0078
⑤	0.006	2.458	0.287	0.245	0.0012	0.003	0.0006	0.0024	0.0081

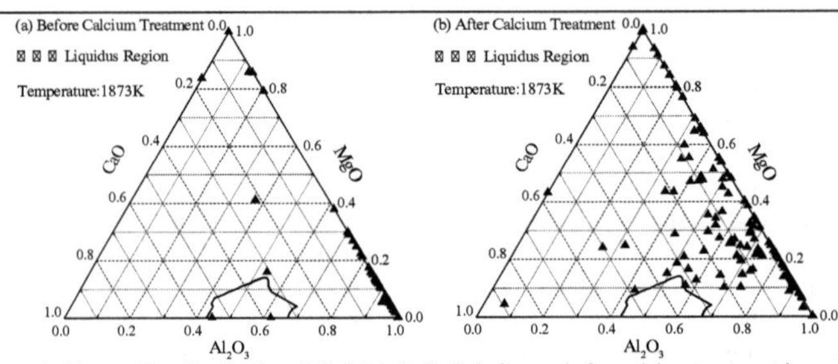

Figure 4. Composition distribution of CaO-MgO-Al_2O_3 before and after calcium treatment in samples

Figure 4 showed the composition distribution of CaO-MgO-Al_2O_3 ternary system in inclusions detected in steel samples before and after calcium treatment. As can be seen in Figure 4, almost all composition points concentrated in Al_2O_3-rich, MgO-low and CaO-few region before calcium treatment, the main kind of inclusions were MgO·Al_2O_3 inclusions, then those data points moved toward the liquidus region (black line in Figure 4) after calcium treatment.

Thermodynamic calculation on the stability diagram of MgO·Al_2O_3 inclusion

In order to obtain the stability diagrams of Mg-Al-O system, thermodynamic calculations were carried out to study the formation of MgO, Al_2O_3, MgO·Al_2O_3 in molten steel.

In Mg-Al-O system, $\log f_{Al}$, $\log f_{Mg}$ and $\log f_O$ were calculated by using the equation of Wagner when both the first-order and the second-order activity interaction coefficients were taken into consideration.

$$\log f_{Al} = e_{Al}^{Al}[\%Al] + e_{Al}^{Mg}[\%Mg] + e_{Al}^{O}[\%O] + r_{Al}^{O}[\%O]^2 + r_{Al}^{Al,O}[\%Al][\%O] + r_{Al}^{Mg,O}[\%Mg][\%O] \quad (1)$$

$$\log f_{Mg} = e_{Mg}^{Mg}[\%Mg] + e_{Mg}^{Al}[\%Al] + e_{Mg}^{O}[\%O] + r_{Mg}^{O}[\%O]^2 + r_{Mg}^{Mg,O}[\%Mg][\%O] + r_{Mg}^{Al,O}[\%Al][\%O] \quad (2)$$

$$\begin{aligned}\log f_O &= e_O^O[\%O] + e_O^{Al}[\%Al] + e_O^{Mg}[\%Mg] + r_O^{Al}[\%Al]^2 + r_O^{Mg}[\%Mg]^2 + r_O^{Al,O}[\%Al][\%O] \\ &\quad + r_O^{Mg,O}[\%Mg][\%O] + r_O^{Al,Mg}[\%Al][\%Mg]\end{aligned} \quad (3)$$

For the formation of MgO, the reaction is [3,16]

$$(MgO) = [Mg] + [O] \quad (4)$$

$$\log K = -4.28 - 4700/T \quad (5)$$

$$\begin{aligned}\log K &= \log\left(\alpha_{Mg}\cdot\alpha_O/\alpha_{MgO}\right) = \log\left(f_{Mg}[\%Mg]\cdot f_O[\%O]/\alpha_{MgO}\right) \\ &= \log f_{Mg} + \log f_O + \log[\%Mg] + \log[\%O] - \log\alpha_{MgO}\end{aligned} \quad (6)$$

For the formation of Al_2O_3, the reaction is [3,16]

$$(Al_2O_3) = 2[Al] + 3[O] \quad (7)$$

$$\log K = 11.62 - 45300/T \quad (8)$$

$$\begin{aligned}\log K &= \log\left(\alpha_{Al}^2\cdot\alpha_O^3/\alpha_{Al_2O_3}\right) = \log\left(f_{Al}^2[\%Al]^2\cdot f_O^3[\%O]^3/\alpha_{Al_2O_3}\right) \\ &= 2\log f_{Al} + 3\log f_O + 2\log[\%Al] + 3\log[\%O] - \log\alpha_{Al_2O_3}\end{aligned} \quad (9)$$

For the formation of MgO·Al_2O_3, the reaction is [3,16]

$$(MgO\cdot Al_2O_3) = 2[Al] + [Mg] + 4[O] \quad (10)$$

$$\log K = 6.736 - 51083.2/T \quad (11)$$

$$\begin{aligned}\log K &= \log\left(\alpha_{Al}^2\cdot\alpha_{Mg}\cdot\alpha_O^4/\alpha_{MgO\cdot Al_2O_3}\right) = \log\left(f_{Al}^2[\%Al]^2\cdot f_{Mg}[\%Mg]\cdot f_O^4[\%O]^4/\alpha_{MgO\cdot Al_2O_3}\right) \\ &= 2\log f_{Al} + \log f_{Mg} + 4\log f_O + 2\log[\%Al] + \log[\%Mg] + 4\log[\%O] - \log\alpha_{Al_2O_3}\end{aligned} \quad (12)$$

The first-order and second-order interaction coefficients were reported by Itoh [9]. The stability diagrams of Al-Mg-O system were calculated by using Matlab and were shown in Figure 5. The calculation (Figure 5(a)) agrees well with that by Itoh [9] and by Yang [2]. As can be seen in Figure 5(b), the MgO·Al_2O_3 region becomes narrower when the activity of MgO·Al_2O_3 is smaller than unity. Meanwhile, the stable region of MgO·Al_2O_3 is also affected by the dissolved oxygen in molten steel. As can be seen, the real composition spots were located in the region of MgO·Al_2O_3 after aluminum addition.

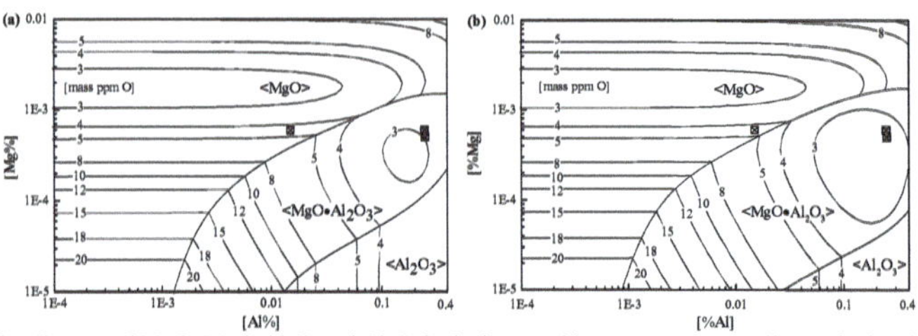

Figure 5 stability diagram of MgO, $MgO\cdot Al_2O_3$and Al_2O_3 inclusions, and iso-oxygen contour lines calculated at 1873K with experimental data point, (a) unity activity of $MgO\cdot Al_2O_3$, (b) 0.5 activity of $MgO\cdot Al_2O_3$.

Size, Morphology and Type of $MgO\cdot Al_2O_3$ Inclusions in Steel Samples

The size, morphology and types of $MgO\cdot Al_2O_3$ inclusions detected in steel samples were analysed by using SEM-EDS and were shown in Figure 6.

It was found in Figure 6(a)-(d) that the observed inclusions were mainly $MgO\cdot Al_2O_3$ spinels before calcium treatment. Moreover, there were some complex inclusions in the form of a thin AlN layer which formed on the surface of an $MgO\cdot Al_2O_3$ spinel inclusion, as shown in Figure 5(c) and (d). These findings suggested that AlN nucleated around $MgO\cdot Al_2O_3$ spinel during its formation.

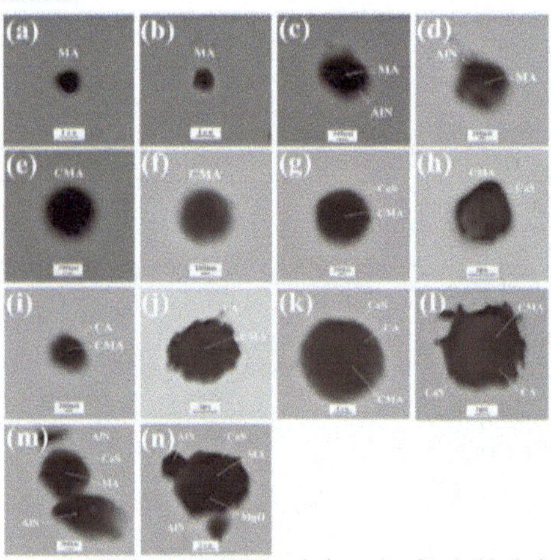

Figure 6. SEM images and EDS analysis results of typical inclusions (MA, CMA and CA represents $MgO\cdot Al_2O_3$, $CaO\cdot MgO\cdot Al_2O_3$ and $CaO\cdot Al_2O_3$, respectively)

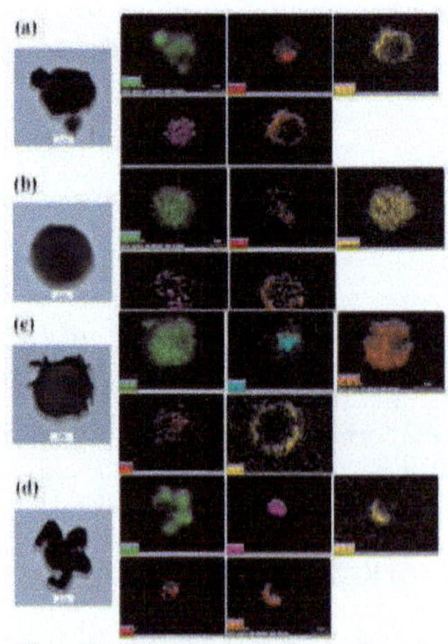

Figure 7. SEM-mappings of main elements (include Al, Ca, Mg, O and S) in typical inclusions

Most of the observed $MgO\cdot Al_2O_3$ spinels by SEM-EDS were irregular or near-spherical with angularities in shape and about 1 m in size. The typical $MgO\cdot Al_2O_3$ inclusions modified after calcium treatment were shown in Figure 6(e) through (n). The majority of inclusions were about 1-2 m in size and regular in shape. Furthermore, inclusions were a little larger than those without calcium treatment.

Figure 6(e) through (h) showed that the main inclusions were calcium aluminates with magnesium oxide ($CaO\cdot MgO\cdot Al_2O_3$). According to SEM images and EDS analysis results, concentration of Al, Mg or Ca elements from the outer layer to the inner part wasn't uniform, it appeared a different distribution in compositions.

Another typical inclusion observed after calcium treatment in samples was shown in Figure 6(i) through (l). It can be seen that, $CaO \cdot MgO \cdot Al_2O_3$ inclusions formed and then were wrapped by the formation of $CaO \cdot Al_2O_3$ inclusions. $CaO \cdot MgO \cdot Al_2O_3$ existed in the center of inclusions, but size, position and morphology differed from each other as shown in Figure 6(i) through (l). Besides the inclusion in Figure 6(i), the other three inclusions were bigger than 3 m in size. Moreover, all the inclusions were spherical or near-spherical as shown in Figure 6(i) through (l). Figure 6(k) and (l) showed that a ringlike CaS inclusion formed around the original inclusion after calcium treatment.

Figure 6(m) and (n) showed a typical inclusion which had an $MgO \cdot Al_2O_3$ spinel core and a shell of CaS inclusion. The inclusion was 1 m in size and 3 m in size, respectively. Furthermore, these two inclusions were near-spherical and this attributed to the formation of CaS inclusion on the surface of $MgO \cdot Al_2O_3$ spinel. As can be seen in Figure 6(n), it existed another small MgO core inside the $MgO \cdot Al_2O_3$ spinel. It was considered that evolution mechanism of this unmodified $MgO \cdot Al_2O_3$ spinel transformed from MgO to $MgO \cdot Al_2O_3$ spinel with spare MgO inclusions before calcium treatment and it supported the evolution model from Park [4].

<u>Elemental Mapping and Line scanning of Observed Inclusions</u>

Figure 7 showed the SEM-mappings of typical inclusions detected in steel samples. It was helpful to figure out the distribution of main elements including Al, Mg, Ca, O and S. It also revealed the distribution region and the change of relative content of each compound which included Al_2O_3, MgO, CaO and CaS in nature. Meanwhile, the mechanism of modification was partially implied by these SEM-mappings.

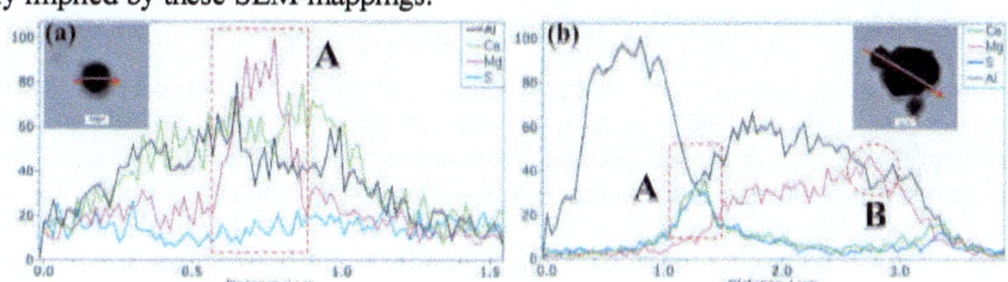

Figure 8. SEM line scanning of main elements (include Al, Ca, Mg and S) in inclusions

Figure 8 showed the SEM line scanning results of typical inclusions observed in steel samples. It revealed the relative content of main elements consisting of magnesium, aluminium, calcium and sulfur. Rectangle A in Figure 8 (a) showed the relative content between magnesium and aluminium in the core of the inclusion, it indirectly confirmed the evolution route of $MgO \rightarrow MgO \cdot Al_2O_3 \rightarrow CaO \cdot MgO \cdot Al_2O_3$. Rectangle A in Figure 8 (b) showed that there was a layer of CaS inclusions around the original $MgO \cdot Al_2O_3$ spinels without modification. Ring B in Figure 8 (b) showed that the content of magnesium element was higher than that of aluminium element in a local region.

<u>Evolution Mechanism of $MgO \cdot Al_2O_3$ Spinel Inclusions During Calcium Treatment</u>

Although some evolution process of $MgO \cdot Al_2O_3$ spinel inclusions had been simply mentioned above, further understanding on evolution mechanism of inclusions was still necessary because this process was also a problem of kinetics. Based on the experimental observations and the preceding discussions, the modification mechanism and behaviors of $MgO{\cdot}Al_2O_3$ spinel inclusions could be summarized by the following steps and the schematic in Figure 10.

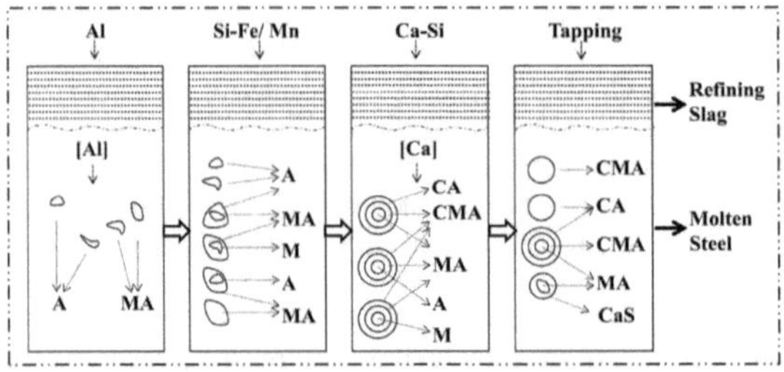

Figure 10. Illustration of evolution of inclusions (M, A and C represents MgO, Al_2O_3 and CaO, respectively)

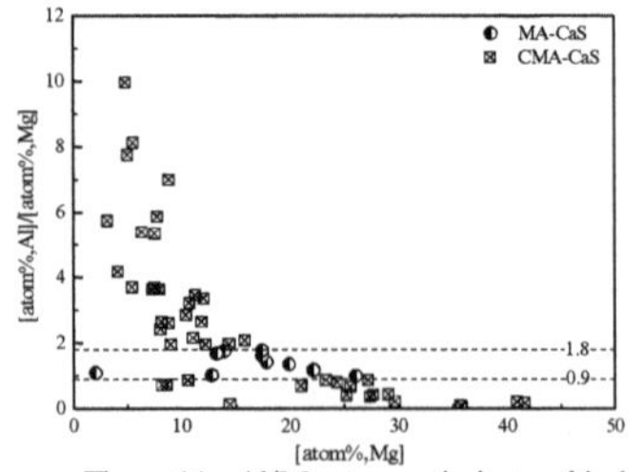

Figure 11. Al/Mg atom ratio in two kinds of CaS bearing inclusions

First step: Formation of MgO Al_2O_3 spinel inclusions before calcium treatment.

With the addition of aluminium element, the transformation of inclusion from Al_2O_3 to $MgO{\cdot}Al_2O_3$ spinel or from MgO to $MgO{\cdot}Al_2O_3$ spinel was illustrated in Figure 10.

Route A: change of inclusion from Al_2O_3 to $MgO{\cdot}Al_2O_3$ spinel [1,3]

$$Mg(1) = Mg^{2+} + 2e^- \quad (13)$$

$$3[Mg]+4(Al_2O_3)=3(MgO\cdot Al_2O_3)+2[Al] \quad \text{or} \quad 3Mg^{2+}+4(Al_2O_3)=3(MgO\cdot Al_2O_3)+2Al^{3+} \quad (14)$$

Route B: change of inclusion from MgO to $MgO{\cdot}Al_2O_3$ spinel [1,4]

$$Al(1)=Al^{3+}+3e^- \quad (15)$$

$$2[Al]+4(MgO)=(MgO\cdot Al_2O_3)+3[Mg] \quad \text{or} \quad 2Al^{3+}+4(MgO)=(MgO\cdot Al_2O_3)+3Mg^{2+} \quad \text{or}$$

$$2Al^{3+}+[MgO_6]^{10-}=(MgO\cdot Al_2O_3)+2O^{2-} \quad (16)$$

$MgO{\cdot}Al_2O_3$ spinel inclusions from these two routes were called MgO-spinel and Al_2O_3-spinel. [1,3-4] At the beginning, Mg atom or Al atom diffused from the bulk into Al_2O_3/molten steel surface or MgO/molten steel surface and was oxidized to Mg^{2+} ions or Al^{3+} ions. Mg^{2+} ions or Al^{3+} ions diffused through the intermediate layer with the diffusion of equimolar vacancies to opposite direction, then reaction (14) or (16) occurred and a thin $MgO{\cdot}Al_2O_3$ spinel formed around the original core. Mg^{2+} ions or Al^{3+} ions continued to diffuse into and through the newly product layer and reacted with Al_2O_3 or MgO. But the reaction rate would decrease with the rise of thickness of the product layer. Of course, pure $MgO{\cdot}Al_2O_3$ spinel inclusions would form under certain conditions. Meanwhile, Al_2O_3 inclusions might form at the surface of the original $MgO{\cdot}Al_2O_3$ spinel inclusions as shown in Figure 6(c) and (d).

Second step: Modification of MgO Al_2O_3 spinel inclusions after calcium treatment.

The number of Mg^{2+} ions in the inclusion replaced by Ca^{2+} ions decided the modification degree of calcium treatment as shown in Figure 10. Reaction steps were given as follows [1,3-4]:

$$[Ca]+[O]=(CaO) \quad \log K=3.292+7220/T \quad \text{or} \quad Ca(1)=Ca^{2+}+2e^- \quad (17)$$

$$xCa^{2+}+(yMgO\cdot zAl_2O_3)=(xCaO\cdot(y-x)MgO\cdot zAl_2O_3)+xMg^{2+} \quad (18)$$

$$Ca^{2+}+(MgO\cdot Al_2O_3)=(CaO\cdot Al_2O_3)+Mg^{2+} \quad \text{or}$$

$$2Ca^{2+}+2[AlO_6]^{9-}=Ca[AlO_6]^{7-}+Ca[AlO_4]^{3-}+2O^{2-} \quad (19)$$

In this step, Ca atom diffused from the steel melt into $MgO{\cdot}Al_2O_3$/molten steel surface and was oxidized to Ca^{2+} ions, then Ca^{2+} ions diffused through the intermediate layer with the diffusion of equimolar vacancies to opposite direction and Mg^{2+} ions in the $MgO \cdot Al_2O_3$ spinel inclusion were replaced by Ca^{2+} ions.

This replacement reaction (18) happened quickly at the interface of reaction. A thin product layer of calcium aluminates with a certain amount of MgO ($CaO \cdot MgO \cdot Al_2O_3$) generated at the outside of the $MgO \cdot Al_2O_3$ spinel inclusion. So the original core of $MgO \cdot Al_2O_3$ spinel became smaller and smaller and the product layer got thicker and thicker.

With the calcium treatment continuing, the rate of reaction (18) became slower and slower due to the decreasing diffusion rate of Mg^{2+} ions in the product layer of calcium aluminates. Then the reaction (19) would dominate the modification process and a possible new liquid product layer of $xCaO \cdot yAl_2O_3$ would also form in the outside of the inclusion in the same time, it would change the original irregular shape into a spherical one due to the surface tension. Of course, if the reaction (18) or (19) lasted long, the original inclusions would transform into pure $CaO \cdot MgO \cdot Al_2O_3$ or $CaO \cdot Al_2O_3$ as shown in Figure 10. But this study hardly observed pure $CaO \cdot Al_2O_3$ without MgO in samples after calcium treatment, it attributed to the large inclusion size, the short refining time and less enough content of dissolved calcium in molten steel.

Effect of the Al/Mg Atom Ratio on the Formation of CaS Bearing Inclusions

Besides the formation of CaS on the surface of calcium aluminates with or without MgO, if the local sulfur content in molten steel was high enough, the reaction (20) would occur [2] and CaS would directly form at the surface of non-modified $MgO \cdot Al_2O_3$ spinel inclusions as shown in Figure 6(m) and Figure 10.

$$[Ca]+[S]=(CaS) \tag{20}$$

$$\log K=-6.485+28340/T \tag{21}$$

As mentioned above, it was needed to consider whether high concentration of sulfur content was the only reason to lead to the direct formation of CaS at the outside of the original $MgO \cdot Al_2O_3$ spinel inclusions before the reaction (18) happened.

Figure 11 showed the Al/Mg atom ratio in two kinds of CaS bearing inclusions which included $MgO \cdot Al_2O_3$-CaS (MA-CaS) and $CaO \cdot MgO \cdot Al_2O_3$-CaS (CMA-CaS). As can be seen, the Al/Mg atom ratio in most of MA-CaS concentrated in a specific region from 0.9 to 1.8, but the Al/Mg atom ratio in CMA-CaS distributed in other regions. It was considered that the Al/Mg atom ratio in the original $MgO \cdot Al_2O_3$ spinel inclusion was one of the reasons to decide whether the dissolved calcium would react with $MgO \cdot Al_2O_3$ spinel to form calcium aluminates and then generate CaS or the dissolved calcium would directly react with sulfur to form CaS in certain conditions.

As mentioned above, it revealed that the Al/Mg atom ratio in a certain range would prevent the diffusion of Ca^{2+} ions and Mg^{2+} ions in inclusion layers, and promote the reaction between calcium and sulfur on the surface of inclusions. Further study on how it works should be paid more attention.

Conclusions

Characteristics of $MgO \cdot Al_2O_3$ spinel inclusions in calcium treated non-oriented silicon steel were studied by laboratory experiments. The evolution and modification mechanism of $MgO \cdot Al_2O_3$ spinel inclusions were also investigated in the current study. The following conclusions can be obtained:

(1) Before calcium treatment, MgO-Al_2O_3 system inclusions detected in steel samples existed four forms. After calcium treatment, CaO-MgO-Al_2O_3 system inclusions also included three

kinds.

(2) The modification degree of $MgO·Al_2O_3$ spinel inclusions which transferred into partially or fully modifying calcium aluminates even with a core of original $MgO·Al_2O_3$ spinel attributed to the incomplete or complete reduction of MgO from $MgO·Al_2O_3$ spinel inclusions by dissolved calcium.

(3) Al/Mg atom ratio had a significant effect on the modification of $MgO·Al_2O_3$ spinel and the formation of CaS, it could decide whether (a) the dissolved calcium started to react with $MgO·Al_2O_3$ spinel and generated calcium aluminates, or (b) the dissolved calcium directly reacted with the dissolved sulfur and form CaS inclusions.

Acknowledgements

The authors express their thanks to national natural science foundation of China for their kind financial support (no. 51274032).

References

[1] M. Jiang, X. H. Wang, B. Chen, and W. J. Wang, "Laboratory study on evolution mechanisms of non-metallic inclusions in high strength alloyed steel refined by high basicity slag", *ISIJ Int.*, 50(2010), 95-104.

[2] S. F. Yang, Q. Q. Wang, L. F. Zhang, J. S. Li, and K. Peaslee, "Formation and Modification of $MgO·Al_2O_3$-Based Inclusions in Alloy Steels", *Metall. Mater. Trans. B*, 43(2012), 731-750.

[3] J. H. Park, D. S. Kim, and S. B. Lee, "Inclusion control of ferritic stainless steel by aluminum deoxidation and calcium treatment", *Metall. Mater. Trans. B*, 36(2005), 67-73.

[4] J. H. Park, and D. S. Kim, "Effect of $CaO-Al_2O_3-MgO$ slags on the formation of $MgO-Al_2O_3$ inclusions in ferritic stainless steel", *Metall. Mater. Trans. B*, 36(2005), 495-502.

[5] Z.Y. Deng, and M.Y. Zhu, "Evolution mechanism of non-metallic inclusions in Al-killed alloyed steel during secondary refining process", *ISIJ Int.*, 53(2013), 450-458.

[6] J. Guo, S. Cheng, and Z. Cheng, "Mechanism of Non-metallic Inclusion Formation and Modification and Their Deformation during Compact Strip Production (CSP) Process for Aluminum-Killed Steel", *ISIJ Int.*, 53(2013), 2142-2151.

[7] E. B. Pretorius, H. G Oltmann, and T. CASH, "The effective modification of spinel inclusions by Ca treatment in LCAK steel", *Iron Steel Technol.*, 7(2010), 31-44.

[8] N. Verma, P. Pistorius, R. Fruehan, M. Potter, H. Oltmann, and E. Pretopius, "Calcium modification of spinel inclusions in aluminum-killed steel: reaction steps", *Metall. Mater. Trans. B*, 43(2012), 830-840.

[9] H. Itoh, M. Hino, and S. Ban-Ya, "Thermodynamics on the formation of spinel nonmetallic inclusion in liquid steel", *Metall. Mater. Trans. B*, 28(1997), 953-956.

[10] K. Fujii, T. Nagasaka, and M. Hino, "Activities of the constituents in spinel solid solution and free energies of formation of MgO, $MgO·Al_2O_3$", *ISIJ Int.*, 40(2000), 1059-1066.

[11] D. Yang, X. Wang, G Yang, P. Wei, and J. He, "Inclusion Evolution and Estimation during Secondary Refining in Calcium Treated Aluminum Killed Steels", *Steel Res. Int.*, 85(2014), 1517-1524.

[12] Y. Kang, F. Li, K. Morita, and S. Du, "Mechanism study on the formation of liquid calcium aluminate inclusion from $MgO·Al_2O_3$ Spinel", *Steel Res. Int.*, 77(2006), 785-792.

2016 EPD Congress

SYMPOSIUM: MATERIALS PROCESSING FUNDAMENTALS

Forming, Joining, Sensing: Devices and Applications

Cong Wang,
Northeastern University

Jonghyun Lee,
University of Massachusetts

EPD Congress 2016
Edited by: Antoine Allanore, Laura Bartlett, Cong Wang, Lifeng Zhang, and Jonghyun Lee
TMS (The Minerals, Metals & Materials Society), 2016

EVALUATION OF JOINT PERFORMANCE ON HIGH NITROGEN STAINLESS STEEL WHICH IS EXPECTED TO HAVE HIGHER ALLERGY RESISTANCE

Kouichi Nakano[1]

[1]Graduate School of Life Science and Systems Engineering, Kyushu Institute of Technology, Hibikino 2-4, Wakamatsu-ku, Kitakyushu-shi, Fukuoka, 808-0196, Japan

Keywords: High Nitrogen Stainless Steel, Austenitic stainless steel, Rare Metal, Resource Saving, Nickel, Anti Metallic Allergy, Stud Welding, Vickers hardness

Abstract

Austenitic stainless steel, which includes nickel for stabilizing austenitic structure, is used for various purposes, for example, for structural material, corrosion-resistant material, biomaterial etc. Nickel is set as one of the rare metals and economizing on nickel as the natural resources is required. On the other hand, nickel is one of the metals that cause metallic allergy frequently. Therefore, high nitrogen stainless steel, where nitrogen stabilizes austenitic structure instead of nickel, has been developed in Japan and some of the foreign countries for the above reason. When high nitrogen stainless steel is fused and bonded, dissolved nitrogen is released to the atmospheric area, and some of the material properties will change. In this study, we bonded high nitrogen stainless steel by stud welding process, which is able to bond at short time, and we evaluate joint performance. We have got some interesting results from the other tests and examinations.

Introduction

Nickel stabilizes austenitic structure and austenitic stainless steel added nickel is used for various purposes, for example, for structural material, corrosion-resistant material, and biomaterial etc. Nickel is one of the rare metals. Economizing of nickel is required as the natural resources. On the other hand, nickel is one of the metals that cause metallic allergy frequently. Therefore, high nitrogen stainless steel, where nitrogen stabilizes austenitic structure instead of nickel, has been developed by solid-state absorption or pressure type electroslag remelting methods in Japan and some of the foreign countries for the above reason. But in case high nitrogen stainless steel is fused and bonded, dissolved nitrogen is released to the atmospheric area, and some of the material properties will change.

In this study, we bonded high nitrogen stainless steel by stud welding process, which is able to bond at short time, and we evaluated joint performance by tensile test, bending test, macro examination according to ASME Sec. IX. The measurement of vertical stability between stud and base plate, micro examination, Vickers hardness test, and Electron Probe Micro Analysis (EPMA) were also carried out.

Materials used and experimental methods

Materials used

SUS304N2 was used as stud material ($\Phi 9 \times L100$ mm), and a tip ($\Phi 0.9 \times L0.9$ mm) was prepared on the head of stud. Nitrogen content of SUS304N2 is 0.20wt%. SUS304N2-X was used as base plate (w30 × L30 × t9.0 mm). Nitrogen content of SUS304N2-X is 0.22wt%. Chemical composition of SUS304N2 and SUS304N2-X are listed in the Table I and II respectively.

Table I Chemical composition of SUS304N2

		Chemical Composition (wt %)								
		C	Si	Mn	P	S	Ni	Cr	N	Nb
Standard requirements	Max.	0.08	1.00	2.50	0.045	0.030	10.50	20.00	0.30	0.15
	Min.						7.50	18.00	0.15	
Measured value		0.04	0.66	1.70	0.030	<0.001	8.22	18.34	0.20	0.08

Table II Chemical composition of SUS304N2-X

		Chemical Composition (wt %)								
		C	Si	Mn	P	S	Ni	Cr	N	Nb
Standard requirements	Max.	0.08	1.00	2.50	0.045	0.030	10.50	20.00	0.30	0.15
	Min.						7.50	18.00	0.15	
Measured value		0.06	0.72	1.85	0.028	0.001	7.69	18.54	0.22	0.10

Stud welding condition

Welding conditions like primary welding voltage (E_0 = 100V), fixing strength (F = 63.7N), electric capacitor (C = 0.117F), and welding time (t = 0.003s) are set constantly. Secondly voltage (welding voltage) was set 130V, 140V, 150V, 160V, 170V and 180V. Materials were cleaned with acetone before welding. Spatter preventing solution was applied for the base plate surface. Spatter preventing solution is mixture of water and surfactant.

The measurement of vertical stability between stud and base plate

The measurement of vertical stability between stud and base plate was based on the welded surface of base plate, and the measurement error was within an angle of 0.1 degree. The measured value was plotted on the circular graph and investigated.

Tensile test

Tensile test was carried out for every five specimens, which were welded at each welding voltage, according to ASME Sec.IX QW-192.1.1. Afterward, every specimen was evaluated weather the value of tensile strength was higher than the least tensile strength of 210MPa or not according to the requirement of ASME Sec.IX QW-192.1.3.

Bending test

Every five specimens, which were welded at each welding voltage, have been bent at 15 degree according to ASME Sec. IX QW-192.1.1. And then, each specimen has been evaluated whether it had cracks or not after returning 15 degree according to the requirement of ASME Sec. IX QW-192.1.1.

Macro and micro examination

Macro examination observing at 10-fold magnification was carried out according to ASME Sec. IX QW-192.1.4. Afterwards, each specimen has been evaluated whether it has cracks or not.
Micro examination observing at 400-fold magnification was carried out for base plate, heat affected zone (HAZ), bond zone, and weld metal of each specimen.

Vickers hardness test
Vickers hardness test was carried out at 0.2 mm intervals from the bond interface to the both outer sides up to 2.0mm in length. Test force of 1.961N and holding time of 5 seconds were set constantly.

EPMA
EPMA was carried out for each cross section of the specimen in an axial direction of the stud weld. Analyzed chemical elements were Ni, Cr, N, C, and O.

Experimental Results and Discussion

Appearance of weld
Figure 1 shows the appearance of stud weld when the spatter preventing solution was applied on the base plate surface (a) and not applied (b). In case the spatter preventing solution was applied, the appearance of the stud weld became good as the ferrule was used on the joint. Melted metal didn't fly away and covered weld, therefore we thought stability of the weld increased in this case.

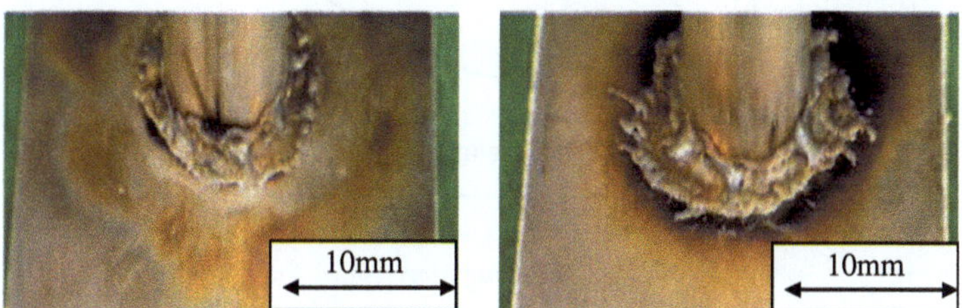

(a) Spatter preventing solution was applied
(b) Not applied

Figure 1. Appearance of the stud welds

Tensile test
Figure 2 shows the relation between welding voltage and tensile strength. All specimens fractured at weld, but satisfied the least tensile strength of 210MPa, which is required by ASME Sec. IX QW-192.1.3.
Figure 3 shows the relation between breaking elongation and tensile strength. Breaking elongation becomes longer according to the increase of tensile strength. Positive correlation between breaking elongation and tensile strength is confirmed.
Figure 4 shows the evaluation model on stud welded joint strength based on droplet behaviour of molten metal. In case of parallel model, tensile strength will increase according to the droplet number of molten metal. On the other hand, in case of serial model, breaking elongation will increase according to the droplet number by this model.
The slope of the positive correlation between breaking elongation and tensile strength changed according to the welding voltage actually. So we think that the mixture ratio of the parallel and serial sequence will change according to the welding voltage.

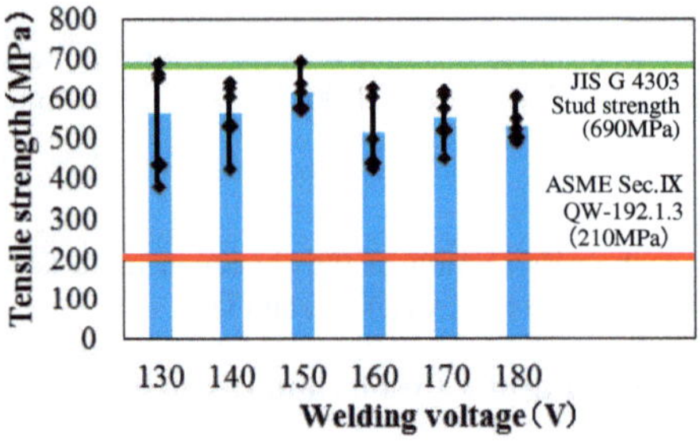

Figure 2. Relation between welding voltage and tensile strength

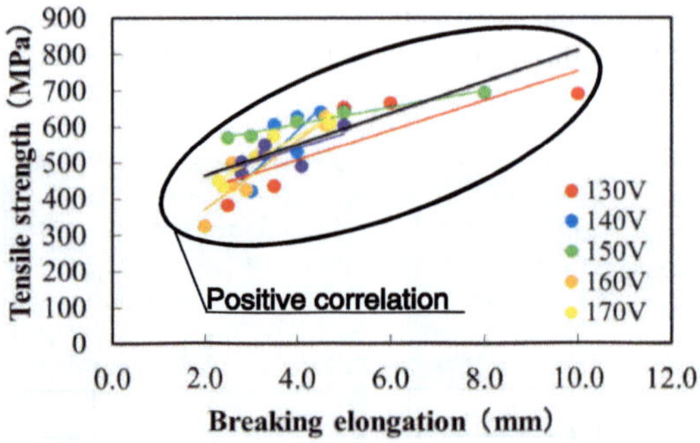

Figure 3. Relation between breaking elongation and tensile strength

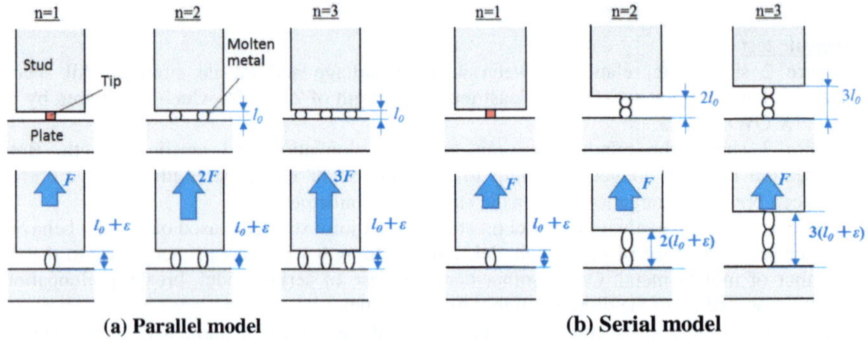

Figure 4. Evaluation model on stud welded joint strength based on droplet behaviour of molten metal (n: number of droplet, F: tensile strength, l_0: original length of droplet, ε : elongation of droplet)

Bending test

All specimens fractured at welds during bending test, and none of the specimen satisfied the requirement of ASME Sec. IX QW-192.1.2 in case welding voltage was 130V and 170V. One specimen satisfied the requirement in case welding voltage was 140V and 160V. Two specimens satisfied the requirement in case welding voltage was 150V and 180V. There is no positive correlation between bending performance and welding voltage. Figure 5 shows the vertical distribution of studs. The maximum value of tilt angle is almost 0.9 degree. So these results have a very little influence on bending performance.

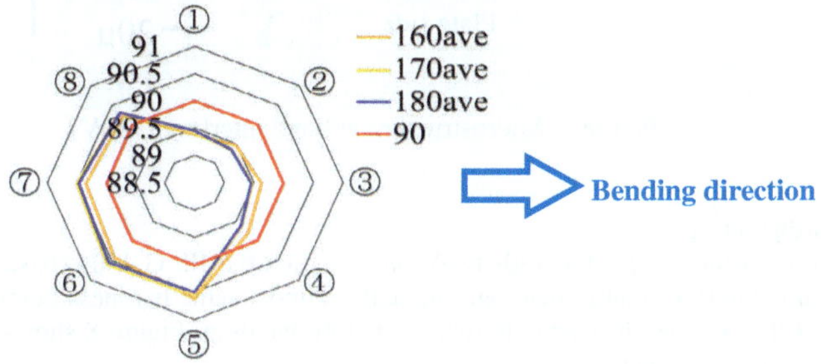

Figure 5. Vertical distribution of studs

Macro and Micro examination

Figure 6 shows the macrograph of cross section of the specimen (160-180V). Some of the lack of fusion area has been observed at the bond interface. Figure 7 shows microstructure at bond interface (140V). We observed dendrite growth toward the welding current direction from the bond interface.

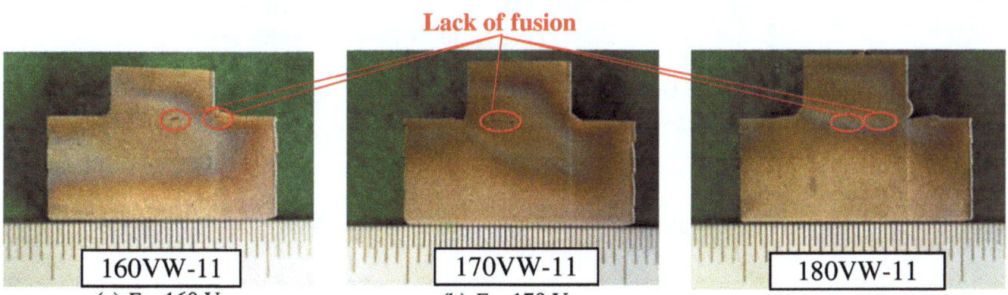

(a) $E = 160$ V (b) $E = 170$ V (c) $E = 180$ V

Figure 6. Macrograph of cross section of specimen (160-180V)

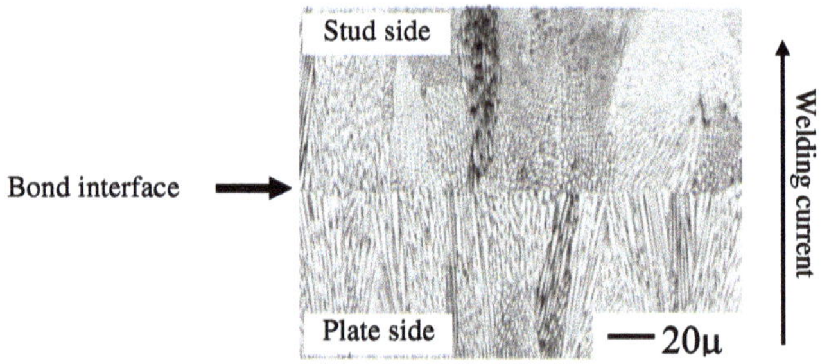

Figure 7. Microstructure at bond interface (140V)

Vickers hardness test
Each specimen which welded at 130~180V has satisfied the JIS G 4304 (base plate) and the JIS G 4303 (stud material) requirement, and indicates almost same hardness before welding. So we think that each specimen has same toughness before welding. Figure 8 shows Vickers hardness distribution from bond interface.

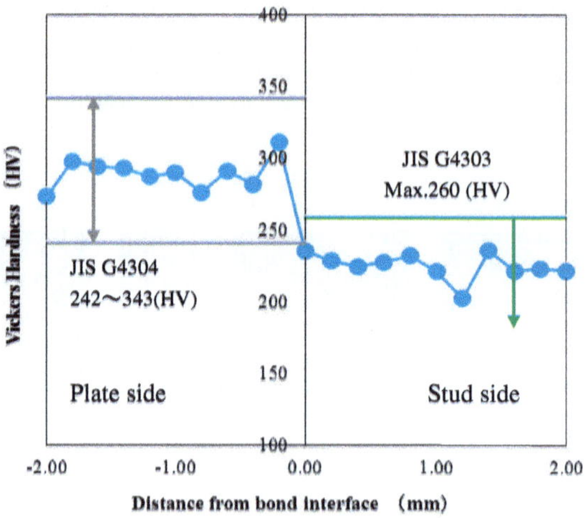

(a) $E = 130$V

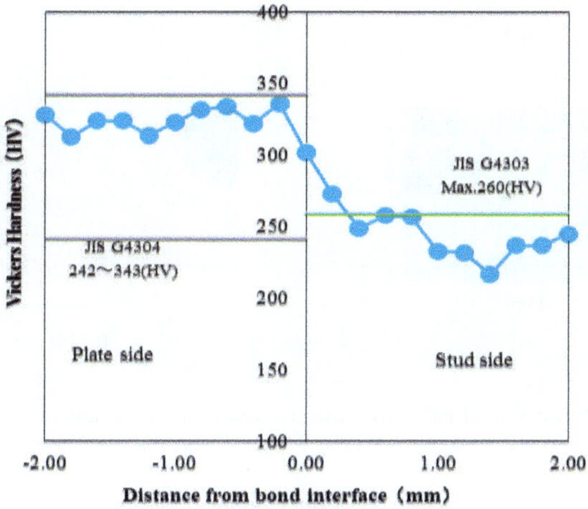

(b) $E = 180\text{V}$

Figure 8. Vickers hardness distribution from bond interface

EPMA

EPMA was carried out at the cross section of specimens. Nitrogen content of the weld metal indicated high enough level. We think that the joint performance will increase if there are no defects in the weldments. The difference of nickel contents between stud and base plate are observed clearly. Figure 9 shows the results of EPMA on the cross section of the specimen in case that the welding voltage is 150V.

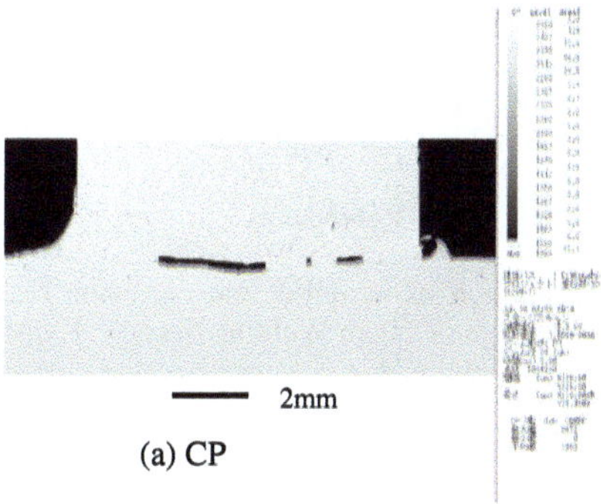

(a) CP

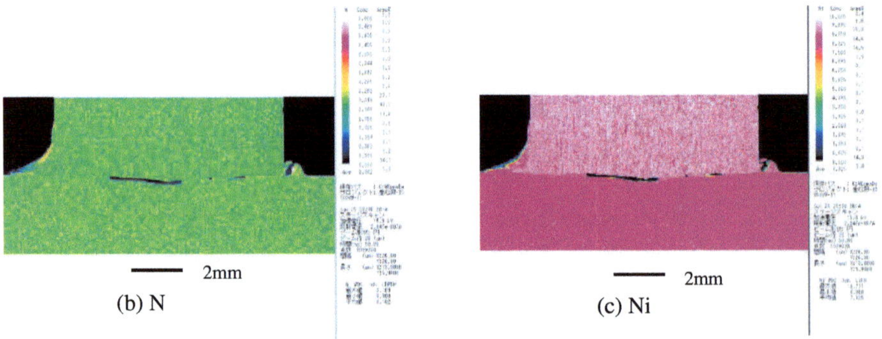

Figure 9. Results of EPMA on the cross section of the specimen (E=150V)

Conclusions

In this study, we bonded high nitrogen stainless steel by stud welding process, which is able to bond at short time, and we evaluated joint performance by tensile test, bending test, macro examination according to ASME Sec. IX. The measurement of vertical stability between stud and base plate, micro examination, Vickers hardness test, and Electron Probe Micro Analysis (EPMA) also carried out. The conclusions obtained in this study are shown as followed.

(1) Weld metal didn't fly away and covered weld, therefore the stability of the weld increased in case of spatter preventing solution was applied.

(2) All specimens satisfied the least tensile strength of 210MPa, which is the requirement of ASME Sec. IX QW-192.1.3. Positive correlation between breaking elongation and tensile strength is confirmed. The slope of the positive correlation between breaking elongation and tensile strength changed according to the welding voltage actually. So we think that the mixture ratio of the parallel and serial sequence will change according to the welding voltage.

(3) There is no positive correlation between bending performance and welding voltage.

(4) Some of the lack of fusion area has been observed at the bond interface. Dendrite growth toward the welding current direction from the bond interface was also observed.

(5) Each specimen has held same toughness before welding.

(6) Nitrogen content of the weld metal indicated high enough level.

References

[1] Kouichi Nakano et al., Evaluation of hardness and corrosion resistance on high nitrogen stainless steel weldments, Proceedings of Japan Welding Society, Kyushu-chapter, No.7, (2010), 41-44 (in Japanese)

2016 EPD Congress

SYMPOSIUM: MATERIALS PROCESSING FUNDAMENTALS

Poster Session

EPD Congress 2016
Edited by: Antoine Allanore, Laura Bartlett, Cong Wang, Lifeng Zhang, and Jonghyun Lee
TMS (The Minerals, Metals & Materials Society), 2016

OBTAINING MULTIPLE METALS THROUGH ELECTRON BEAM MELTING OF REFRACTORY METAL WASTES

Katia Vutova[1*], Vania Vassileva[1]

[1]Institute of electronics, Bulgarian Academy of Sciences;
72 Tzarigradsko shosse; 1784 Sofia, Bulgaria
*e-mail: katia@van-computers.com

Keywords: electron beam melting, refining, tungsten, impurities, wastes

Abstract

Investigations and results on the refining of tungsten scrap applying electron beam melting (EBM) in vacuum are presented and discussed. In this work EB melting experiments were performed in single or double-melt operations with different power inputs and refining times for tungsten purification and recovery of multiple metals (including refractory and other metals such as Mo, Nb, Cu, Zn, etc.), which are alloyed elements or impurities with high concentrations in the initial materials. Evaluations for extraction of some valuable metals, generated in the condensate at their removal from the liquid metal during the refining process are provided. Efficient technological regimes for e-beam melting and refining that enable the simultaneous production of pure metal ingots and condensates that can be used directly or can be used for subsequent extraction of expensive and valuable metals are presented.

Introduction

The recycling of scrap, containing refractory metals such as tungsten, molybdenum, vanadium, tantalum or metals which are reactive (chemically-active at high temperatures) such as titanium, hafnium, zirconium and their alloys is a question of current interest due to: (i) the uniqueness of their characteristics and their multiple applications in different branches of science and engineering, such as metallurgy, energetics, chemical industry, electronics, space industry, etc.; (ii) limited raw material resources; (iii) necessity for expensive equipment for their production. These metals are strategic resources and their value is high. Wastes, containing refractory and reactive metals are important secondary raw material resources for the production of these metals. The majority of the metal wastes contain high quality metal that is polluted or oxidized while being processed.

Electron beam melting (EBM) method has shown advantages in comparison to other metallurgical methods. Recently, EBM has become a widely used technology [1-14]. There are no specific requirements concerning the initial material and thus EBM can be successfully used for processing of material wastes – metals and their alloys.

In EBM refining, processes take place mainly at the interface liquid metal/vacuum in three zones: the front part of the feeding rod (initial material), the drops falling toward the liquid metal pool, and the liquid metal pool in the water-cooled crucible [3, 13, 15]. In addition, refining processes take place in the front of the feeding rod before the temperature reaches the melting temperature value. Reactions also take place during crystallization.

In each reaction zone and at the interface of two or three phases, heterogeneous reactions simultaneously take place which overall rate depends on multiple phenomenon: the matter velocity towards the interface surface, the chemical reactions that take place at this interface, the removal rate from the surface, the heat transfer conditions, etc. [9, 16]. The rate of some reactions such as reduction, degassing, evaporation of volatile components, etc. can become so low that they limit the overall refining process [6, 17-19].

Investigation are carried out and experimental data and results for EBM use for recycling of tungsten wastes with high concentrations of other valuable metals such as molybdenum, niobium, cobalt, copper, etc. are presented in the paper. Chemical analysis data and refining processes at different technological regimes are discussed. Metallographic analysis of samples at single and double EB melting and before EBM is also made. Extraction of some important metals, generated in the condensate is investigated and it is shown that the efficiency of the EB recycling of wastes with high level of metal impurities can be increased.

1. Experimental

The experiments were carried out in the laboratory "Physical problems of electron beam technologies", Institute of electronics – BAS by using an ELIT-60 electron beam installation with capacity 60 kW and accelerating voltage of 24 kV. The EB melting furnace used in the study was composed of the melting chamber, optical system (one electron gun), vacuum system, and feeding (horizontal) and extraction system (the drip molten metal crystallized in the water-cooled copper crucible). The vacuum system was made of double sets of rotary pump, oil-diffusion pump for maintaining an operation pressure in the chamber about $5\text{-}8\times10^{-3}$ Pa and one turbo molecular pump was set for the e-beam gun.

Table I. Melting conditions of EBM in the experiments with tungsten scrap materials M1, M2, and M3

raw material	M1	M2		M3	
	single EBM	*single (EB1) and double (EB2) melting*		*double EBM*	
e-beam power P_b [kW]	10, 13, 15 kW	single melting $EB_{1,1}$	12 kW	17.5 kW	
		double melting (in two operations $EB_{1,2}$ and $EB_{2,1}$)	4.8 kW for $EB_{1,2}$ and 7.2 kW for $EB_{2,1}$		
melting time τ [min]	15 min	single melting $EB_{1,1}$	15 min	first double melting	60 min
		double melting (in two operations $EB_{1,2}$ and $EB_{2,1}$)	5 min for $EB_{1,2}$ and 5 min for $EB_{2,1}$	second double melting	180 min

EB melting experiments were performed in single (EB_1) or double-melt operations (the EB_1 ingot was remelted by EB_2) for different tungsten wastes (with different concentrations of tungsten and impurities). Three types of tungsten scrap materials (M1, M2, M3) were used as raw materials. The first investigated scrap material (M1) contains 88 % tungsten and molybdenum, niobium, cobalt and nickel with high concentrations; the second one (M2) contains

25 % tungsten in the form of WC, 45 % Fe, molybdenum, cobalt, etc.; the third scrap material contains 76.69 % tungsten and high quantity of copper and zinc.

The influence of the e-beam power and the melting time on the removal efficiency and on the material losses is investigated in order to increase the efficiency of the e-beam recycling of every investigated scrap material. The values of the experimental parameters electron beam power (P_b) and melting time (τ) are shown in Table I.

2. Results and Discussion

2.1. Tungsten Scrap Material M1

For the first investigated tungsten scrap material (M1) the experiments were performed upon the single EBM for three e-beam powers – 10, 13 and 15 kW and the melting time was 15 min (Table I). Chemical analysis was performed by using emission spectral analysis before and after being refined by single EBM (Table II). For every investigated technological regime the overall removal efficiency η, [%] (removal efficiency of all the impurities of the sample), the removal efficiency of a specific impurity (η_{imp}, [%]) and the material losses (ΔG, [%]) due to evaporation are estimated (Table II).

At refining of tungsten wastes M1 with high level of Mo and Nb, it is found that, the increase of the e-beam power leads to an increase of the removal efficiency of these impurities (η_{Mo}, η_{Nb}). Maximal removal efficiency of Mo and Nb, e.g. minimal concentration of Mo (C_{Mo},[%]) and Nb (C_{Nb}, [%]) are seen at P_b=15 kW (Table II, where C_i is the overall concentration of the all other impurities of the material except for Mo and Nb, G_0 is the weight of the initial ingot).

Table II. Impurities' concentration, removal efficiency and material losses at EBM of tungsten scrap M1

	C_W [%]	C_{Mo} [%]	C_{Nb} [%]	C_i [%]	η [%]	G_0 [g]	ΔG [%]	G_{cond} [g]	$C_{Mo,cond}$ [%]	$C_{Nb,cond}$ [%]
before EBM	88.02	0.3	0.25	11.43						
after EBM $P_b = 10$kW	98.97	0.25	0.2	0.58	91.4	301.4	28.23	85.1	0.427	0.377
after EBM $P_b = 13$kW	99.02	0.25	0.2	0.53	91.82	315.9	28.43	89.8	0.426	0.376
after EBM $P_b = 15$kW	99.2	0.2	0.19	0.41	93.32	418.7	32.72	137.0	0.51	0.374

The variation of the removal efficiency η and the increase of the material losses ΔG at an increase of the e-beam power (overheating of the liquid pool, respectively) is due to changes in the concentrations of the other impurities such as Ni, Co, Cr, etc. (Table II) with lower content in the initial material M1. Evaluations for extraction of some valuable metals (such as Mo, Nb), generated in the condensate at their removal from the liquid metal during the refining process are made. The concentrations of Mo ($C_{Mo,cond}$, [%]) and Nb ($C_{Nb,cond}$, [%]) in the condensate for every investigated regime are also presented in Table II.

The obtained results show that EBM of tungsten wastes with high level of Mo, Nb, Co, Ni (material M1) leads to impurities removal and high efficiency removal η > 91 % can be achieved. The increase of the beam power leads to an increase of the efficiency removal and the tungsten ingot with purity of 99.2 % is obtained from the raw material M1 by virtue of single EBM (Table II). The evaluations made show that up to 6 g refractory metals (such as Mo, Nb) can be also obtained by processing of the condensate.

2.2. Tungsten Scrap Material M2, Containing 25 % W in the Form of WC and High Level of Fe, Co and Mo

The raw material M2 was processed by single (EB_1) and double EBM (EB_2) e-beam melting and different technological regimes (conditions) were realized. The values of the process parameters are:

- for the single $EB_{1,1}$ melting: P_b = 12 kW and melting time τ = 15 min;
- for the single $EB_{1,2}$ melting: P_b = 4.8 kW, τ = 5 min; for the double $EB_{2,1}$ melting (the ingot through $EB_{1,2}$ melting is processed by double $EB_{2,1}$ melting): P_b =7.2 kW, τ =5 min.

Data about chemical analysis of the impurities' concentration of the starting material M2 (before EBM) and of the ingots after EBM of the tungsten scrap is obtained and analyzed. In Table III data about inclusions' concentration after single $EB_{1,1}$ melting and after double $EB_{2,1}$ refining, material losses (ΔG) and removal efficiency (η, η_{Fe}, η_{Mo}, η_{Co}) is presented, C_i is the overall concentration of the all other impurities of the material except for Fe.

Table III. Impurities' concentration, removal efficiency and material losses at single ($EB_{1,1}$) and double ($EB_{2,1}$) melting of tungsten scrap M2

	C_W [%]	C_{Fe} [%]	η_{Fe} [%]	C_{Co} [%]	η_{Co} [%]	C_{Mo} [%]	η_{Mo} [%]	C_i [%]	η [%]	ΔG [%]	$C_{Co,cond}$ [%]	$C_{Mo,cond}$ [%]
before EBM	25	45		3.5		1.1		30				
single $EB_{1,1}$ melting	96.18	3.5	92.2	0.1	97.1	0.004	99.6	0.32	95	4.86	70	22.5
double $EB_{2,1}$ melting	96.74	3.0	93.3	0.1	97.1	0.004	99.6	0.26	96	7.57	45	14.5

For the tungsten scrap recycling from waste product M2, the more efficient refining process is seen at double e-beam melting $EB_{2,1}$ and a high purity tungsten ingot is obtained (Table III). The processes at double EBM are more efficient in regard to shorter melting time and lower e-beam power, and high removal efficiency (η =96 %) is seen. The value of the removal efficiency of the major impurity Fe (η_{Fe}) of the material being refined by single ($EB_{1,1}$) and double ($EB_{2,1}$) e-beam melting is also high (92-93 %) and does not depend on the technological conditions realized for M2 material recycling. The overall concentration of the other impurities such as Mn, Mg, Ti, Ca, Al, V, Cu, Zn, Ni, Cr, Sb, S, etc. (except for the selected important impurities Co and Mo) of the raw material M2 is 0.4 % and the concentration of each of them after $EB_{1,1}$ and $EB_{2,1}$ melting is in the range of 10 - 300 ppm and also does not depend on the technological regime.

Practically complete extraction of the valuable metals Co and Mo for both technological regimes used is seen and removal efficiency η_{Co} and η_{Mo} after $EB_{1,1}$ and $EB_{2,1}$ are 97.1 % and

99.6 %, respectively (Table III). The cobalt quantity generated in the condensate is more than a half of the condensate mass and the molybdenum mass is about a quarter of the obtained condensate (Table III). Thus, from the processed condensate about 700 g Co and 225 g Mo can be obtained as well as W ingot with more than 96 % purity is produced by the virtue of EBM.

2.3. Tungsten Scrap Material M3, Containing 76.69 % W and High Level of Cu, Zn and Co

The raw material was preliminary oil removed, dried and compacted in the form of disks. The purity of the obtained in this way initial material M3 was W 76.69 % and contained high levels of Cu, Zn and Co impurities (Table IV). The EB melting experiments were performed in double-melt operations (the EB_1 ingot was remelted by EB_2) for e-beam power P_b= 17.5 kW and different melting times: the overall refining time was 60 min and 180 min for the investigated cases, respectively (Table I).

Table IV. Tungsten and impurities concentration before EBM and after double e-beam melting (EB_2)

	C_W [%]	C_{Cu} [%[	C_{Zn} [%]	C_{Co} [%]	C_{Sn} [%]	C_{Ta} [%]	C_i [%]
before EBM	76.69	11	10	1.25	0.35	0.3	0.41
double EBM $\tau = 60$min	99.4	0.04	0.001	0.117	0.0017	0.03	0.4103
double EBM $\tau = 180$min	99.38	0.03	0.001	0.15	0.0017	0.03	0.4073

The concentration of Cu (C_{Cu}), Zn (C_{Zn}), Co (C_{Co}), Sn (C_{Sn}) and Ta (C_{Ta}) impurities as well as the overall concentration of all the other impurities (C_i) of the material before EBM and after being refined by double EB melting are shown in Table IV. It was found that the recycling of the tungsten wastes M3 with high level of Cu, Zn and Co inclusions was a relatively slow process. For these experiments enough time for effective refining process is 60 min (Table IV), at which the best purification of W (99.4) and maximal removal efficiency of the inclusions (such as Zn, Sn, Co, Ta, Cr and Ti) for which there are not thermodynamics limitations for their removal at EBM are obtained.

Metallographic analysis of samples at both double EB melting (for τ= 60min and τ= 180min) and before EBM is made and the structures of tungsten samples are shown in Fig.1. In the matrix tungsten before EBM presence of two other structure compositions – phases of copper and zinc is seen. As seen in Table IV, the major impurities Cu and Zn are removed significantly after double EB melting for τ = 60min and structure with well-formed tungsten grains is seen (Fig.1b). The longer refining time τ = 180min at double EB melting does not lead to significant impurities' removal (Table IV). The additional refining time mainly leads to structure improvement - redistribution of the impurities in the matrix tungsten and formation of perfect structure without defects (Fig.1c).

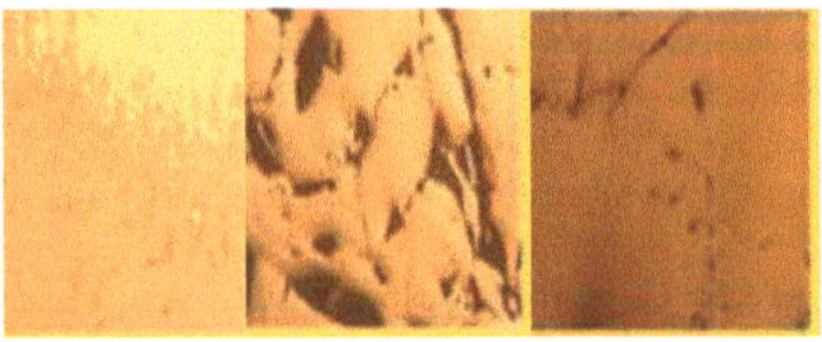

(a) (b) (c)
Figure 1. Structures of tungsten samples: (a) before EBM and after double-melt operations for (b) $\tau = 60$ min, and for (c) $\tau = 180$ min.

The major metallic impurities Cu and Zn of the tungsten scrap material M3 are non-ferrous metals and their concentration in the raw material is Cu 11 % and Zn 10 %. The concentration of these metals in the generated condensate is respectively Cu 37 % and Zn 34 %. The extracted non-ferrous metals are relatively cheap and an additional complex processing of the condensate will add more expenses to the used resources (energy, time, consumables, etc). While recycling such scrap materials the goal for the EBM is the tungsten with good purity (99.4) to be compacted and poured in appropriate form so that it can be mainly used in metallurgy as an alloying component in production of special steels. The efficiency of recycling raw materials with content similar to M3 for obtaining materials with high purity is questionable due to the fact that the technological process is considerably slow.

3. Conclusions

The conclusions of this experiment of recycling of tungsten scrap materials with different impurity contents in the raw materials (M1, M2, M3) by single and double electron beam melting and refining at different technological regimes are as follows:

- For materials (such as M1) with major metallic impurities which are:
  - expensive and valuable metals (such as Mo, Nb);
  - with low concentrations in the raw material (lower than 1%);
  - with insufficient removal efficiency due to thermodynamic limits at the refining conditions which, on the other hand, are optimal conditions for the refining of tungsten,

the efficiency concerning additional processing of the generated condensate in order to extract these impurities (such as Mo, Nb) as pure metals should be preliminary evaluated in regards to their market demand and price.

- For materials (such as M2) with major metallic impurities which:
  - are expensive and valuable metals (such as Co, Mo);
  - are with high concentrations in the raw material (higher than 1%);
  - have more than 97% removal efficiency after single and double EB refining for technological conditions which are also optimal process conditions for the tungsten purification,

the additional processing of the obtained condensate aiming to obtain these impurities (such as Mo, Co) as pure metals is recommended and is justified taking into account their characteristics and applications.

- For materials (such as M3) with major metallic impurities which:
  - are cheap metals (such as Cu, Zn);
  - are with high concentrations in the raw material (higher than 10%);
  - are sufficiently removed with removal efficiency of more than 99 % following the double EBMs at process conditions which at the same time are also optimal conditions for the tungsten purification,

the condensate can be used directly (for example as alloyed elements in metallurgy) - without additional processing that will increase the price of the included non-ferrous metals.

The approach for EBM of scrap materials containing refractory metals depends on the level of their purity. In the paper, efficient technological regimes that enable the simultaneous obtaining pure tungsten ingots and condensates that can be used directly or can be used for subsequent extraction of expensive and valuable metals are presented. The obtained results show that the efficiency of the EB recycling of wastes with high level of metal impurities can be increased – both for obtaining refined refractory metal ingots and for extraction and use of some important metallic impurities removed.

4. Acknowledgements

This research was funded by the National Fund for Scientific Research at the Ministry of Education and Science of Republic of Bulgaria under contract BIn-5/2009. The authors would like to express most sincere gratitude to their Indian colleagues from C-MET, Hyderabad for the useful discussions on the subject.

5. References

1. A. Mitchell and T. Wang, "Electron beam melting technology review," Proc. of the Conf. *Electron Beam Melting and Refining*, State of the Art 2000, Reno, NV, USA, (2000), ed. R. Bakish, 2–13.
2. R. Bakish, *J. Min. Metall. Mater. S.*, 50 (1998), 28–30.
3. G. Mladenov et al., "Experimental and theoretical studies of electron beam melting and refining," in book *Practical Aspects and Applications of Electron Beam Irradiation*, eds. M. Nemtanu and M. Brasoveanu, (Transword Research Network: Trivandrum, India, 2011), 43–93.
4. J.M. Oh et al., "Preparation of ultrahigh purity cylindrical tantalum ingot by electron beam drip melting without sintering process," *Materials Science and Technology (UK)*, 29 (5) (2013), 542-546.
5. J.M. Oh et al., "Preparation and purity evaluation of 5N-grade ruthenium by electron beam melting," *Metall. Trans.*, 53 (2012), 1680–1684.
6. V. Vassileva et al., "Oxygen removal during electron beam drip melting and refining," *Vacuum*, 77 (2005), 429-436.
7. J. Chen et al., "Production of 5N copper by directional solidification and electron beam refining," *Zhenkong Kexue yu Jishu Xuebao/Journal of Vacuum Science and Technology*, 31 (4) (2011), 495-499.
8. L. Murr et al., "Metal fabrication by additive manufacturing using laser and electron beam melting technologies," *J. Mater. Sci. Technol.*, 28 (2012), 1–14.
9. T. Zhang et al., "High-Purity Nickel Prepared by Electron Beam Melting: Purification Mechanism," *Metallurgical and Materials Transactions B*, 45 (1) (2014), 164-174.

10. Y. Tan and S. Shi, "Progress in research and development of electron beam technology in metallurgy refining field," *Cailiao Gongcheng/Journal of Materials Engineering*, 8 (2013), 92-100.
11 Y. Ma et al., "Preparation procedure of high purity tungsten via electron beam side surface melting," *Cailiao Kexue yu Gongyi/Material Science and Technology*, 22 (1) (2014), 30-35.
12. J.-K. Lee et al., "Directional Solidification Behaviors of Polycrystalline Silicon by Electron-Beam Melting," *Japanese Journal of Applied Physics*, 52 (10 PART2) (2013), art. no. 10MB09, http://dx.doi.org/10.7567/ JJAP.52.10MB09.
13. K. Vutova et al., "Investigation of Electron Beam Melting and Refining of Titanium and Tantalum Scrap," *Journal of Materials Processing Technology*, 210 (2010), 1089-1094.
14. Z. Wang et al., "Research progress in production technology of titanium," *Xiyou Jinshu/Chinese Journal of Rare Metals*, 38 (5) (2014), 915- 927.
15. K. Vutova et al., "Application of non-stationary thermal model for simulation and investigation of heat and refining processes of Ti during EBMR," *EPD Congress 2013, TMS (The Minerals, Metals & Materials Society)*, (Wiley, San Antonio, USA, 2013), 253-260.
16. V. Vassileva, K. Vutova, and G. Mladenov, "Analysis of the Thermodynamic Conditions of Refining during Electron Beam Melting of Refractory Metals," *Materials Science and Engineering Technology*, 37 (7) (2006), 613-618.
17. J.M. Oh et al., "Deoxidation of Ti Powder and Preparation of Ti Ingot with Low Oxygen Concentration," *Materials Transactions*, 53 (06) (2012), 1075-1077.
18. K. Vutova et al., "Electron beam melting and recycling of hafnium", *Supplemental Proceedings: v.3: General Paper Selections TMS2011* (Wiley, San Diego, USA, 2011), 725-732.
19. K.-M. Roh et al., "Comparison of deoxidation capability for preparation of low oxygen content powder from TiNi alloy scraps," *Powder Technology*, 253 (2014), 266-269.

EPD Congress 2016
Edited by: Antoine Allanore, Laura Bartlett, Cong Wang, Lifeng Zhang, and Jonghyun Lee
TMS (The Minerals, Metals & Materials Society), 2016

SOLIDIFICATION AND EVALUATION OF THERMAL PARAMETERS OF Sn-Zn EUTECTIC ALLOYS HORIZONTALLY SOLIDIFIED

Alex Iván Kociubczyk[1,2], Roberto Rozicki[2], Verónica Liliana Scheiber[2], Alicia Esther Ares[1,2]

[1]Materials Institute of Misiones, IMAM (CONICET-UNaM), 1552 Azara Street, Posadas, Misiones, 3300 Argentina
[2] Faculty of Sciences, University of Misiones; 1552 Azara Street, Posadas, 3300 Argentina

Keywords: Sn-Zn eutectic alloys, horizontal directional solidification, thermal parameters.

Abstract

In the present research the horizontal solidification of Sn-Zn eutectic alloys (Sn-8.9wt.%Zn) is performed, with two opposite senses of heat extraction. From the process of solidification, the thermal parameters (local solidification time, cooling rates and interphases velocities) and grain sizes from macrostructures obtained were determined. The presence of defects in the solidified samples was analyzed, observing that they occur largely in the collision zones of solidification fronts advancing in opposite directions in the samples.

Introduction

Experiences previously realized under controlled conditions of vertical unidirectional solidification, which evidenced the columnar-to-equiaxed transition (CET), with lower cooling and uncontrolled conditions of unidirectional solidification with superior cooling (convection) show that under certain circumstances four interphases of solidification are formed, resulting in further solidification of internal areas of the samples and consequently hollows, defects and internal voids in the unidirectional solidified samples [1]. In a previous work, horizontal directional solidification of Sn-Zn hypereutectic alloys [2, 3], with equiaxed structures with predominant heat extraction from both ends of the samples were analyzed, which led to six solidification interphases, two liquid interphases or [liquid / (solid + liquid)] (or [L / (S + L)] or I_L), two eutectic interphases or [(solid + liquid) / (solid + liquid /eutectic + liquid)] (or [(S + L) / (E + L)] or I_E) and two solid interphases or [(eutectic + liquid) / solid] (called [(E + L) / S] interphases or I_S), moving in opposite directions [4, 5].
Since it is important in the solidification area the coupling of experimental macroscopic determinations with parameters affecting the solidification process and the fact that only by experimental validation, solidification models can be applied with enough confidence [7-10].
In a previous research, horizontally directionally solidified samples with a predominant heat extraction in both opposite directions were determined, leading to six solidification interphases: two liquid interphases, I_L, two eutectic interphases, I_E, and two solid interphases, I_S, moving in opposite directions.
The experimental determination of the velocities and accelerations of interphases of each pair of interphases present in the process and the determining the functions of these velocities and accelerations thus constitute the first step, necessary to characterize the kinetics of solidification of these alloys under the experimental conditions imposed.
This paper aims to solidify horizontally directionally Sn-Zn eutectic alloys and thus analyze the evolution of the thermal parameters.

Experimental Procedure

Samples of Zn-Sn eutectic alloys were prepared by directional solidification. The Zn and Sn pure elements were merged into graphite molds in a muffle furnace, and then unidirectionally solidified in clay mold in the horizontal furnace with heat extraction in two opposite directions (see Figure 1). Next, the samples were ground with SiC abrasive paper of different grain size, from # 60 to # 1500. To observe the macrostructure, the samples were subjected to a chemical attack which consisted of exposing them to a solution of 36.5% HCl at room temperature between 5 to 30 seconds. This allowed defining zones of different structures along the sample: a columnar zone of large and elongated grains and the equiaxed zone with smaller grains.

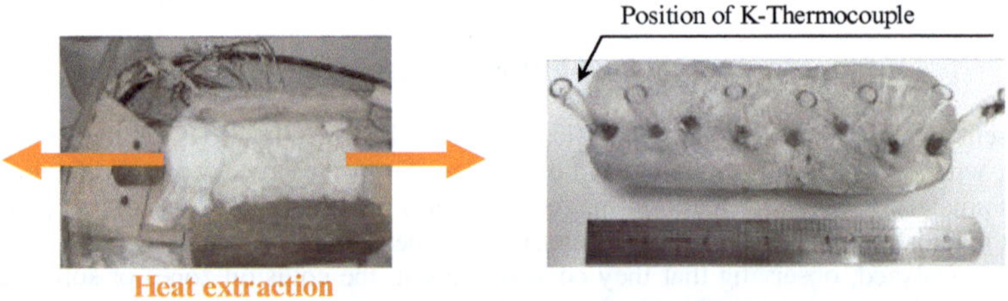

Figure 1. Photographs of experimental device and one sample obtained after horizontal directional solidification process (glass tubes indicate the positions where the thermocouples were placed).

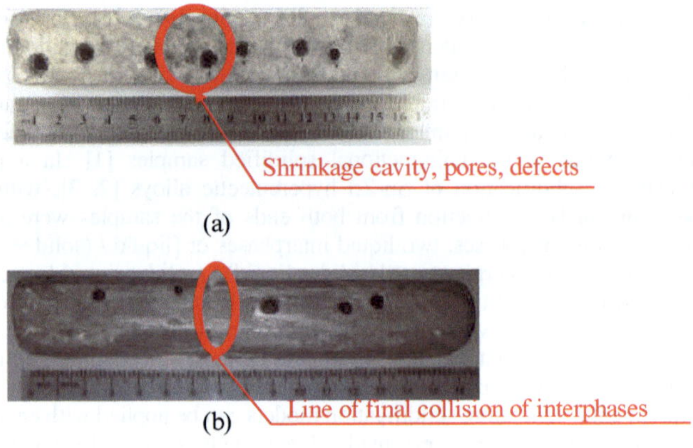

Figure 2. (a) Sample obtained from the horizontal directional solidification process (Sn-8.9% Zn). Eight thermocouples type-k were placed in the positions of the holes. The macrostructure is showing the zone of shrinkage cavity, pores, defects, etc. (b) Macrostructure showing the line of final collision of both interphases advancing in opposite directions and the columnar-to-equiaxed transition (CET).

After the horizontal solidification, metallographic treatment of the samples was measured, for which, one face of each sample was roughened using sandpaper of different grain sizes (between # 60 and # 1500) and then was chemical attacked with HCl (70 %) for about 120 seconds to reveal the macrostructure (see Figure 2). The grain size was measured using ASTM E112 standard.

Results and Discussion

From the collected temperature data, thermal parameters such as, the local solidification time in the positions of each thermocouple, temperature gradients and cooling rates, were determined. Plotting the data of temperature versus time (Figure 3), the presence of three zones on the graph was observed: first - a period of cooling of the melt (temperature overheating and the liquidus temperature); second - solidification period (between the liquidus temperature and the eutectic temperature); third - a period of cooling of the solid (below the eutectic temperature). During the solidification period, it could be noted that the average value of eutectic temperature was 198.5°C.

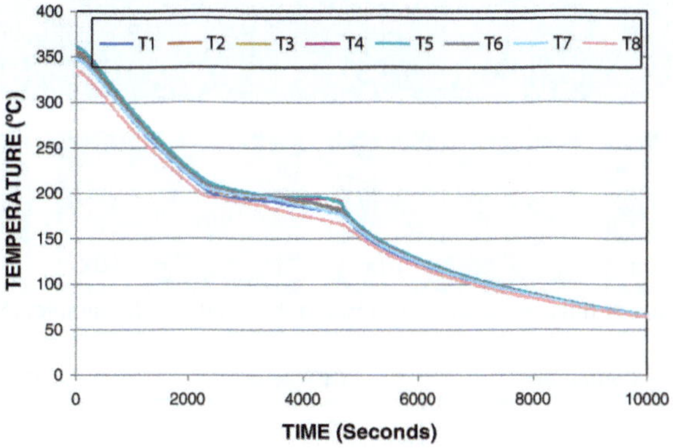

Figure 3. Cooling curves of eutectic alloy (Sn-8.9wt.% Zn). T_1 is the thermocouple located at the left end of the sample and T_8 is at the far right.

In macrographs obtained (Figure 4) the presence of the columnar-to-equiaxed transition (CET) was observed, indicating a high heat removal from the ends of the samples. In one of the experiments carried out the caloric extraction was so high, which induced a completely columnar grain structure. The average size of the equiaxed grains as measured by ASTM E112 standard is presented in Table 1. In general, the larger grain size was obtained at the center of the samples.

Table 1. Grain sizes of equiaxed grains measures in the samples.

Alloy	Grain size, [mm] (Right end)	Grain size, [mm] (Left end)	Grain size, [mm] (Center)
Sn-8.9wt.%Zn	3.9	3.3	5.2

In Figure 4 presents the microstructure of each zone of the sample. Also, in Figure 5 can be seen the evolution of lamella size in the length of the sample. It is observed that the lamella size is smaller at both ends of the sample. Also, the heat extraction was more efficient on the right side.

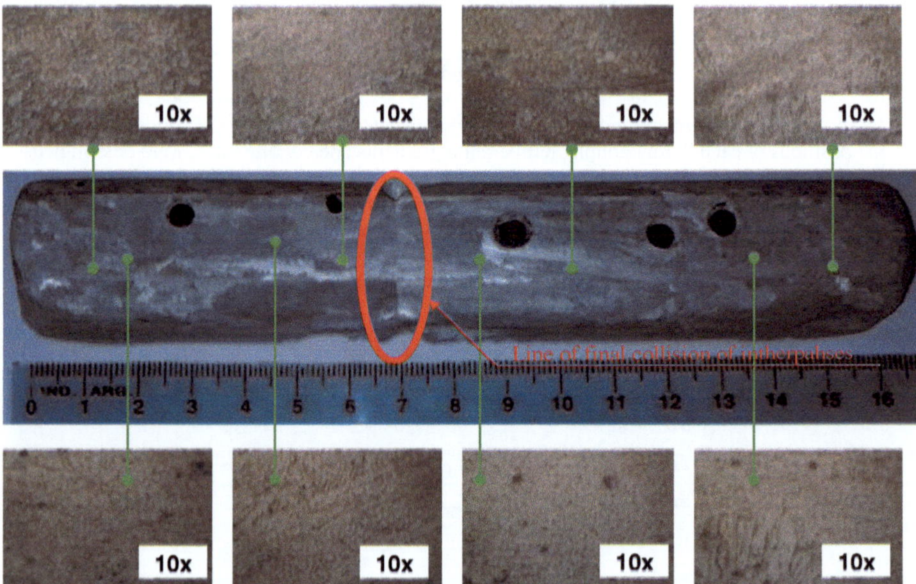

Figure 4. Details of microstructure evolution in the length of the sample. (Sn-8.9wt.% Zn).

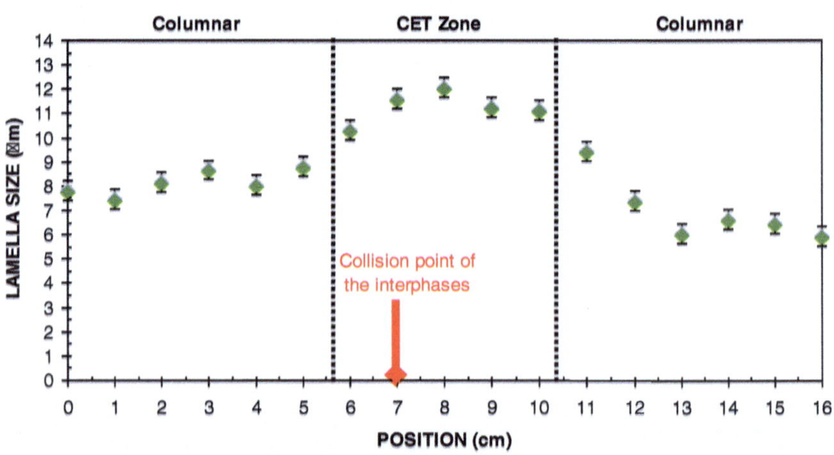

Figure 5. Lamella size evolution versus position along the sample.

Figure 6 shows the position of liquid [L/(L+S)] and solid [(S+L)/S] interphases during solidification of the sample versus time. It can be seen that, although each pair of interphases of the same type is represented by a single line, each line corresponds to two interphases of the same type, since each interphase can not be in two points in the space at the same instant.

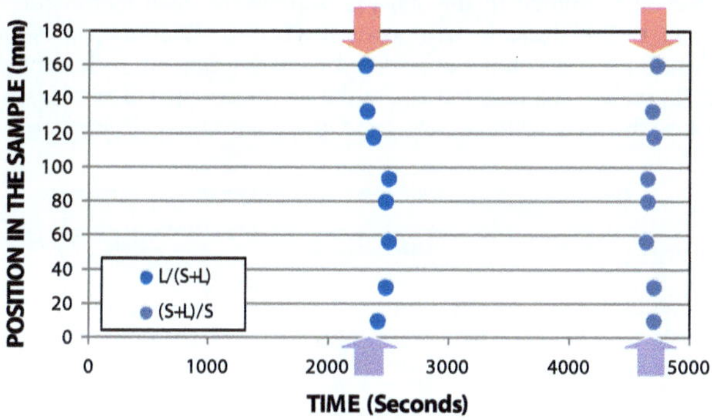

Figure 6. Position of the interphases versus time. Sn-8.9wt.% Zn.

The instant of collision of two interphases of the same type is characterized by the maximum moment in the considered position or further from the curves as point "C". Thus, in the curves of Figure 6 are represented four interphases but only two curves are shown. Each curve corresponds to a pair of interphases advancing in opposite directions and collides at some point inside the sample (pink and violet arrows). Note that the collision points of the interphases or furthermost points of the different curves "C" do not coincide in the same position of the sample, which is the maximum in time vs. position curve (t vs. x). The graph of average position in the sample versus velocity of the liquidus interphase, V_L, is showed in Figure 7. The maximum value in this graph corresponds to the position of collision of both liquidus interphases in the sample.

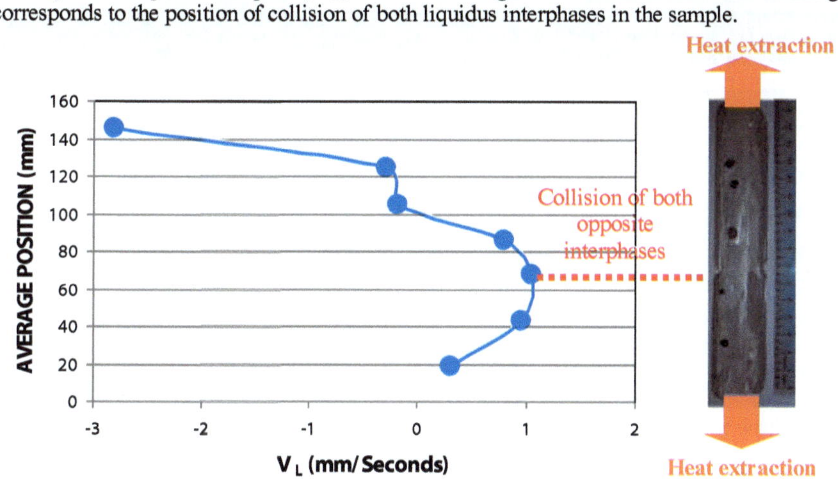

Figure 7. Graph of average position in the sample versus velocity of the liquidus interphase. The negative values of velocities correspond to the interphase advancing in opposite direction.

From temperature versus time data obtained for the alloy concerned, as indicated in Figure 3, the cooling rates were determined in the liquid, mushy and solid alloys taking the derivative values of temperature versus time for each thermocouple position (see Figure 8 for the position of thermocouple T_3). From the resulting changes in the cooling rates the approximate start instants of solidification were determined in the sensing volume for each thermocouple, or instant of liquid interphase, [L/(L+S)], through each position where the thermocouples are. Also, the end instants of solidification in the volume considered were determined, or instant in that each solid interphase, [(E+L)/S], passes the position where the thermocouple is located, see Table 2. In the same Table is showed the local solidification time in the positions of each thermocouple.

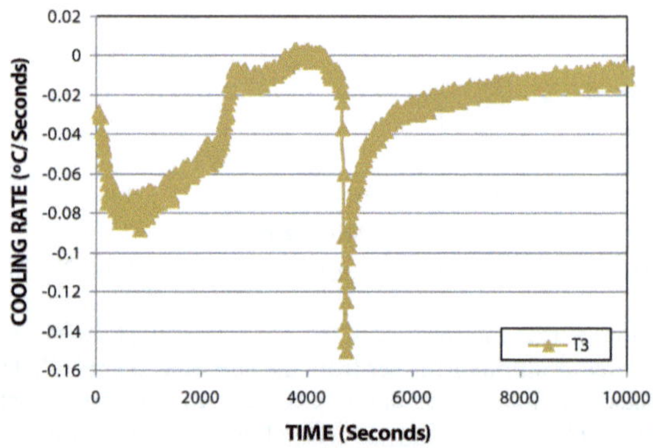

Figure 8. Cooling rate versus time for the position of thermocouple T_3.

Table 2. Values of position of each thermocouple and thermal parameters.

Thermocouple #	Position (mm)	T_L (Seconds)	T_S (Seconds)	T_{SL} (Seconds)	V_L (mm/Seconds)
T1	10	2430	4610	2180	0.32
T2	30	2490	4620	2130	0.97
T3	57	2520	4650	2130	1.05
T4	80	2500	4670	2170	0.8
T5	94	2520	4690	2170	-0.17
T6	118	2390	4720	2330	-0.28
T7	133	2340	4730	2390	-2.8
T8	160	2330	4740	2410	

Since the sample cooled predominantly from both ends, it is possible to determine a pair of interphases for each type of interphase, that is, two liquid and two solid interphases. Each pair of interphases of the same type only moves in the longitudinal direction of the sample but have opposite directions of movement (see Figure 9).

The calculated values of V_L^{-1} expressed as a function of time are shown in Figure 10. The maximum time in the graph corresponds to the time of the collision of two oppositely advancing interphases. This collision causes the presence of voids, pores and internal defects in the samples (areas marked with red circle in macrographs Figures 2 (a) -. (b) and Figure 4.

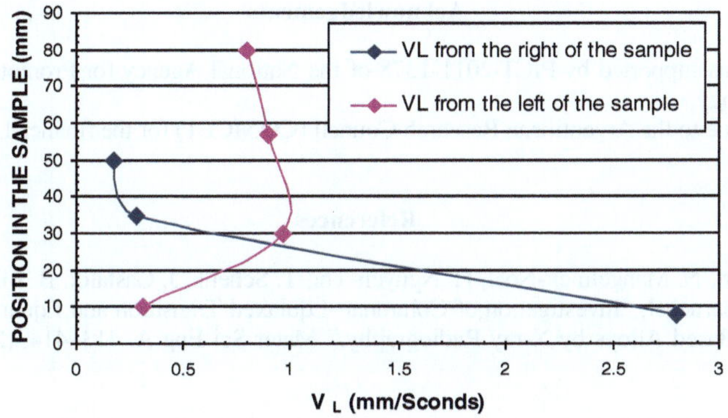

Figure 9. V_L advancing in opposite directions in the sample.

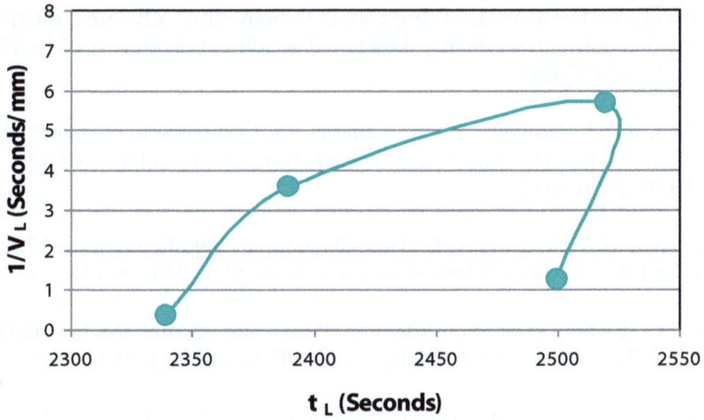

Figure 10. $1/V_L$ versus time (Sn-8.9wt.%Zn).

Conclusions

Horizontal directional solidification experiments were performed with two directions of heat extraction, coincident with the longitudinal axis of Sn-Zn eutectic alloys (Sn-8.9wt.% Zn).
The main parameters were determined, namely: a) the begin and end of solidification at each position considered, b) local solidification time, c) cooling rates, d) the average velocities against advancing liquid interphases in opposite directions e) the value of the velocity when collide solidification fronts, advancing in opposite directions in the samples (liquid interphases collide in a different position than do solid interphases).

Acknowledgements

This work was supported by PICT-2011-1378 of the National Agency for Promotion of Science and Technology.
Thanks are due to the Argentinean Research Council (CONICET) for the financial support.

References

1. G. Reinhart, N. Mangelinck-Noël, H. Nguyen-Thi, T. Schenk, J. Gastaldi, B. Billia, P. Pino, J. Härtwig, J. Baruchel, "Investigation of Columnar–Equiaxed Transition and Equiaxed Growth of Aluminium Based Alloys by X-ray Radiography." Mater Sci Eng A, 413–414, (2005), p. 384–388.

2. W. Desrosin, C. E. Schvezov, A. E. Ares, Thermal and Metallographic Parameters Evolution During Solidification of Zn-Sn Alloys, TMS Annual Meeting (2014) 501-508.

3. A.I. Kociubczyk, W. Desrosin, L.M. Boycho, C.E. Schvezov, A.E. Ares, Interphase evolution and defect formation during horizontally directional solidification process of Sn-Zn alloys, TMS Annual Meeting (2015) 57-64.

4. S. McFadden, D. J. Browne, C. A. Gandin, A Comparison of Columnar-to-Equiaxed Transition Prediction Methods Using Simulation of the Growing Columnar Front. Metall. Mater. Trans. 40A, (2009), p. 662-672.

5. S. F. Gueijman, C. E. Schvezov, A. E. Ares, Directional Solidification and Characterization of Zn-Al and Zn-Ag Diluted Alloys, Materials Transactions, 51, (2010), p. 1851-1870.

6. A. E. Ares, C. E. Schvezov, Solidification Parameters during the Columnar-to-Equiaxed Transition in Lead-Tin Alloys. Metall. Mater. Trans., 31A, (2000), p. 1611-1625.

7. A. E. Ares, C. E. Schvezov, Influence of Solidification Thermal Parameters on the Columnar-to-Equiaxed Transition of Aluminum-Zinc and Zinc-Aluminum Alloys. Metall. Mater. Trans., 38A, (2007), p.1485-1499.

8. S. F. Gueijman, C. E. Schvezov, A. E. Ares, Theory and Experiments of Interphase Formation and Structure in Horizontal Directionally Solidified Zn-Al Dendritic Alloys, In: Crystal Growth: Theory, Mechanisms and Morphology, Chapter 8, Pages: 265-293, Editors: Nicole A. Mancuso and James P. Isaac. Nova Science Publishers, Inc. ISBN: 978-1-61324-529-3 (2012).

9. M. Zurko, C. M. Rodriguez, C. E. Schvezov, C. M. Mendez, A. E. Ares, Horizontal Directional Solidification of Zn-Al Alloys, TMS Annual Meeting (2012) 153-160.

10. C. A. Gandin, "From Constrained to Unconstrained Growth during Directional Solidification". Acta Mater, 48, (2000), p. 2483–2501.

EPD Congress 2016
Edited by: Antoine Allanore, Laura Bartlett, Cong Wang, Lifeng Zhang, and Jonghyun Lee
TMS (The Minerals, Metals & Materials Society), 2016

LASER ABLATION OF ALUMINUM AND TITANIUM ALLOYS UNDER GLASS CONFINEMENT

Peixuan Ouyang[1], Liangju He[2], Peijie Li[1, 3]

[1]Department of Mechanical Engineering, Tsinghua University, Beijing, 100084, China
[2]School of Aerospace, Tsinghua University, Beijing, 100084, China
[3]National Center of Novel Materials for International Research, Tsinghua University, Beijing, 100084, China

Keywords: Laser ablation, Gap width, Laser intensity, Crater profiles, Crater morphologies, Phase explosion.

Abstract

Single-pulse laser ablation of aluminum and titanium alloys under glass confinement is investigated. Processing parameters such as laser intensity and gap width between targets and glass are varied. Surface morphologies and crater profiles of the alloys after laser ablation are studied by SEM and white-light interferometric microscope, respectively. The effects of gap width and laser intensity on surface morphologies and crater profiles of targets are systematically analyzed. In addition, the generation of porous surface structures in Ti target is regarded as the result of phase explosion in the molten surface layer of targets and the thermodynamic condition in our experiment for phase explosion is evaluated.

Introduction

Pulsed-laser ablation of solid substrates under the confinement of transparent glass has shown great potential in laser shock peening (LSP) [1], nanoparticle deposition [2], laser shock cleaning (LSC) [3] and generation of porous metal surface for chemical and electrochemical application [4]. The laser-induced plasma is effectively confined between the target surface and the glass overlay such that strong shock wave is generated acting on the targets. Thus, laser intensity and gap width between target surface and glass greatly affect the shock wave pressure and the processing results of targets. In order to have better understanding and full application of glass-confined laser ablation, the effect of the gap width and the laser intensity on laser ablation process is investigated respectively.

Two modes of laser ablation were used in our work. One is "semiconfined ablation", which is first proposed by Lugomer [4], in which a laser beam irradiates the target through a transparent glass positioned slightly before the target surface. In this experimental configuration, the effect of the gap width on laser ablation of Al alloy is investigated. The other one is "confined ablation", in which no gap exists between the glass and the target. In this experimental configuration, the effect of laser intensity on laser ablation of Al and Ti alloys are studied.

Experimental Procedures

The experimental setup of "semiconfined ablation" and "confined ablation" are shown in Fig. 1a and Fig. 1b, respectively. For the "semiconfined ablation" experiment, aluminum 5A06 alloy (referred to as Al target in the following) was irradiated using a Q-switched Nd:YAG laser (wavelength λ=1064 nm, pulse duration τ=10 ns, intensity I~7.5×10^{10} W/cm^2) at single-pulse mode. The gap width between targets and glass was controlled through transparent tapes with monolayer thickness of 350 m as spacers. For the "confined ablation" experiment, no transparent tapes exist between targets and glass. Al alloy and titanium TB5 alloy (referred to as Ti target in the following) were irradiated by a single laser pulse at combined wavelengths of 1064 nm and 532 nm with total intensity ranging from 7.5×10^{10} W/cm^2 to 3.7×10^{11} W/cm^2. For both semiconfined and confined experiments, soda lime glass slides with thickness of 1 mm were used. Rectangular shaped targets with thickness of 1 mm were grinded and then cleaned with distilled water and acetone before laser ablation.

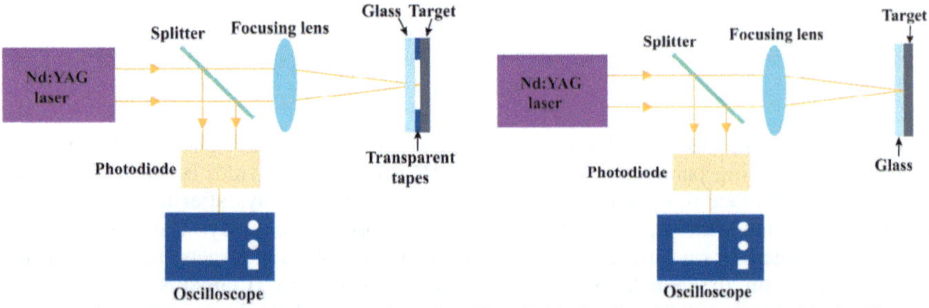

Fig. 1. Experimental setup. (a) "semiconfined ablation"; (b) "confined ablation".

After laser ablation, the targets were cleaned using ultrasound cleaning in acetone for 15 min and in alcohol for 5 min. The surface morphologies of produced craters were investigated using Scanning Election Microscope (Hitachi-S4800). The cross-sectional profiles of craters were measured through white-light interferometric microscope (MicroXAM-3D) with horizontal and vertical resolutions of 0.11-8.8 m and 0.01 nm, respectively. The depth of craters is defined as the distance between the horizon at level zero and the lowest point on the crater profiles. For the case of semiconfined experiment, the back side surfaces of glasses after laser irradiation were observed using optical microscopy.

Results and Discussion

Effect of Gap Width on Laser Ablation of Al Alloy ("Semiconfined Ablation")

Fig. 2 shows surface morphologies of craters in Al target induced by glass-semiconfined ablation with different gap widths. For the crater induced with the gap width of 1.4 mm, re-solidification structure with slight and disordered liquid movement is observed (Figs. 2a and 2d). For the crater induced with the gap width of 700 m, more intensive but still disordered liquid wave is presented (Figs. 2b and 2e). While for the crater induced with the gap width of 350 m, strong radial liquid jets and strings of drops broken from liquid jets due to the instability of jets [4] are formed in the ablated area (Fig. 2c), and slightly porous surface with micro-cavities regarded as bubble nucleation sites [5, 6] appears in the central part of the crater (Fig. 2f). With the gap width decreasing, space becomes very limited for expansion of the ablated plasma, resulting in the formation of high-temperature-high-pressure plasma disk trapped between the target surface and the glass [4]. As a result, bubble nucleation is prone to occur and strong shock wave induced by hot plasma acts on the molten pool of targets, leading to the morphologies shown in Fig. 2c and 2f.

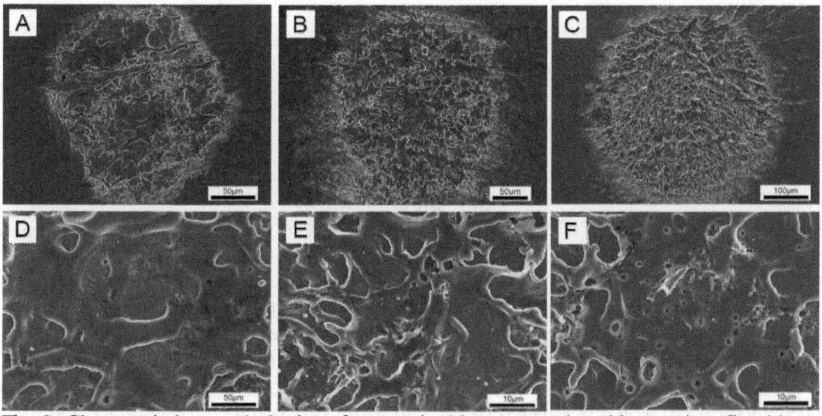

Fig. 2. Characteristic morphologies of craters in Al target produced by semiconfined laser ablation with different gap widths. From left to right, the corresponding gap width is 1.4 mm, 700 m and 350 m. The top row shows overall view of the craters, and the bottom row presents the center of the craters.

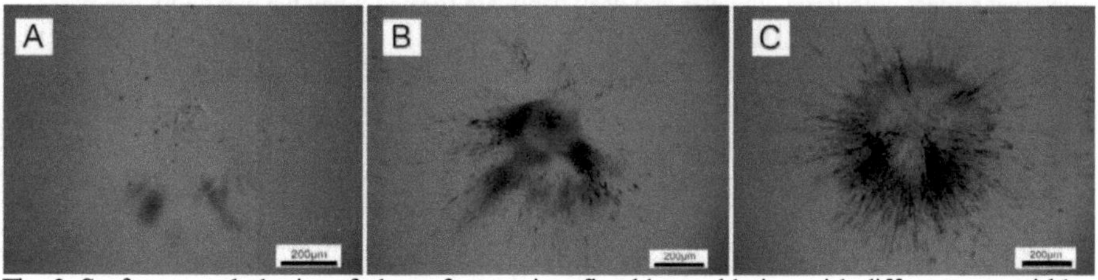

Fig. 3. Surface morphologies of glass after semiconfined laser ablation with different gap widths. (a) 1.4 mm; (b) 700 m; (c) 350 m.

The morphologies of back side surfaces of glasses after semiconfined ablation are shown in Fig. 3. Little damage can be observed in the glass for the ablation experiment with gap width of 1.4 mm (Fig. 3a). With the gap width decreasing, damage in the glass becomes serious. Fig. 3b shows some liquid jets spread from a spot along radius on the glass surface for the experiment with gap width of 700 m, while more radial liquid jets presents on the glass for the experiment with gap width of 350 m (Fig. 3c), which also demonstrates that the shock wave pressure increases with the gap width decreasing.

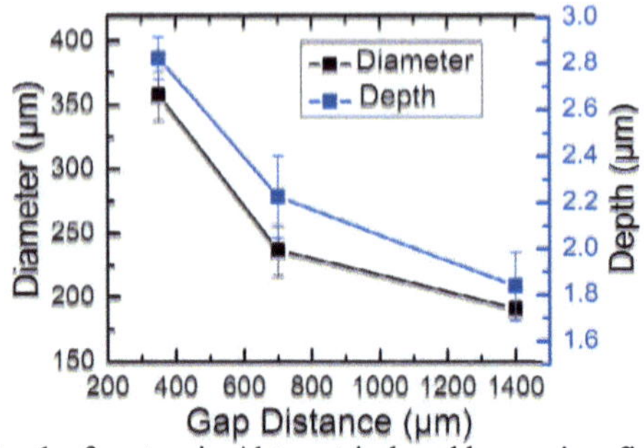

Fig. 4. Diameter and depth of craters in Al target induced by semiconfined laser ablation as a function of gap width.

Fig. 4 presents diameter and depth of the craters in Al target after semiconfined ablation as a function of gap widths. With the gap width increasing, both the diameter and depth of the craters decrease, consistent with coupling coefficient decreasing with gap width increasing [7]. The sizes of crater diameter and depth are closely related to the dynamics and geometry of plasma-induced shock wave which is affected by the gap width between the target surface and the glass [3, 4]. When the gap width decreases from 1.4 mm to 350 m, the shock wave pressure acting on the target increases significantly [3] and the geometry of the shock wave evolves from oblate spheroid to cylindrical slab [4], which implies that the generation of a quasi-two-dimensional shock wave. Subsequently, more materials would be removed from the targets by the shock wave such that craters with large diameter and deep depth are formed when the gap width decreases.

<u>Effect of Laser Intensity on Laser Ablation of Al and Ti Alloys ("Confined Ablation")</u>

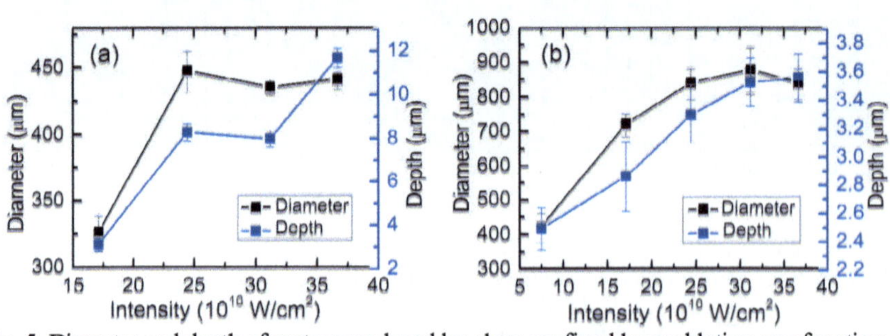

Fig. 5. Diameter and depth of craters produced by glass-confined laser ablation as a function of laser intensity. (a) Al alloy; (b) Ti alloy.

Glass-confined laser ablation of Al and Ti alloys was carried out with the laser intensity ranging from 7.5×10^{10} W/cm^2 to 3.7×10^{11} W/cm^2. Since the crater in Al target induced with laser intensity of 7.5×10^{10} W/cm^2 is unapparent due to the low laser intensity, the diameter and depth of this crater is ignored in the analysis. Fig. 5 demonstrates diameters and depths of craters induced by glass-confined ablation as a function of laser intensity. For Al target, the crater diameter tends to become saturated when the laser intensity is greater than 2.45×10^{11} W/cm^2, but the crater depth still have a rapid increase when the laser intensity ranges from 3.1×10^{11} W/cm^2 to 3.7×10^{11} W/cm^2 even though the crater depth tends to be stable in the range of the laser intensity from 2.45×10^{11} W/cm^2 to 3.1×10^{11} W/cm^2. For Ti target, the crater diameter increases with the laser intensity increasing but slightly decreases when the intensity is greater than 3.1×10^{11} W/cm^2, while the crater depth of Ti target keeps increasing when the laser intensity ranges from 7.5×10^{10} W/cm^2 to 3.7×10^{11} W/cm^2. It is concluded that the crater diameter is easier to reach a saturated value than the crater depth with the laser intensity increasing, which could be explained by that the shock wave is more effective in increasing the depth rather than the width [8]. Consequently, crater depth could be more representative than crater diameter as a function of laser intensity for both Al and Ti alloys.

Nevertheless, it is worth noting that with the same laser intensity and the same spot size, the crater depth of Ti target is much smaller than that of Al target while the crater diameter of Ti target is greatly larger than that of Al target. Since Al alloy has a lower melting temperature and a higher thermal penetration depth ($\delta_T = \sqrt{\alpha t_p}$, where α is the thermal diffusion coefficient and t_p is the pulse width of the laser beam) than Ti target, a thicker molten surface layer is produced in Al target instead of Ti target and a deeper crater would be generated after the recoiled pressure acting on the molten layer. As for the phenomenon of the crater diameters, it is probably because greater shock wave pressure is generated during laser ablation of Ti alloy rather than Al alloy. According to Fabbro's theory [9], the maximum shock wave pressure generated during confined laser ablation is given by the following relation:

$$P(\text{GPa}) = 0.01\sqrt{\frac{\alpha}{\alpha+3}} \times \sqrt{Z(\text{g}\cdot\text{cm}^{-2}\cdot\text{s}^{-1})} \times \sqrt{I_0(\text{GW/cm}^2)}\,, \tag{1}$$

where α is the fraction of internal energy devoted to thermal energy (typically α=0.1), I_0 is the incident power intensity, and Z is the reduced shock impedance between the target and the liquid medium, which is defined by the relation:

$$\frac{2}{Z} = \frac{1}{Z_{\text{glass}}} + \frac{1}{Z_{\text{target}}}\,, \tag{2}$$

where Z_{glass} and Z_{target} are the acoustic impedances of glass and target respectively. Since the acoustical impedance of Al (1.5×10^6 g cm^{-2} s^{-1}) is smaller than that of Ti (2.7×10^6 g cm^{-2} s^{-1}), and the incident laser intensity reaching the surface of Al target is lower than that of Ti target due to the higher reflectance of Al, the maximum shock wave pressure generated in laser ablation of Ti is larger than that of Al, which results in craters with larger diameter in Ti target than that in Al target.

Fig. 6 shows characteristic morphologies of craters in Al target produced by glass-confined ablation with two laser intensities of 1.7×10^{11} W/cm^2 and 3.7×10^{11} W/cm^2. Compared with the crater induced with the laser intensity of 1.7×10^{11} W/cm^2 (Figs. 6a and 6c), the crater generated with the laser intensity of 3.7×10^{11} W/cm^2 presents intensive radial liquid jets moving outwards

(Figs. 6b and 6d), which indicates that a stronger shock wave is generated when laser intensity increases. Fig. 7 presents characteristic morphologies of craters in Ti target after confined laser ablation with three intensities of 7.5×10^{10} W/cm^2, 1.7×10^{11} W/cm^2 and 3.7×10^{11} W/cm^2. From the overall view of these craters (Figs. 7a-7c), it is clear that the crater diameter of Ti target greatly increases with laser intensity increasing. As for the detail view of the center of the craters, more micro-cavities, liquid jets and droplets are formed in the crater when the laser intensity increases from 7.5×10^{10} W/cm^2 to 1.7×10^{11} W/ cm^2 (Figs. 7d and 7e), which is ascribed to the molten layer with high temperature and strong recoiled shock wave due to the increasing incident laser energy. Note that the morphology of the crater generated with the laser intensity of 3.7×10^{11} W/cm^2 (Fig. 7f) is much different from that of the other two craters (Figs. 7d and 7e). The corresponding magnified SEM micrograph is shown in Fig. 8. Smaller cavities are formed inside the larger ones through a series of bubble generation and the cascade of bubble explosion, the phenomenon of which is belonged to phase explosion [4].

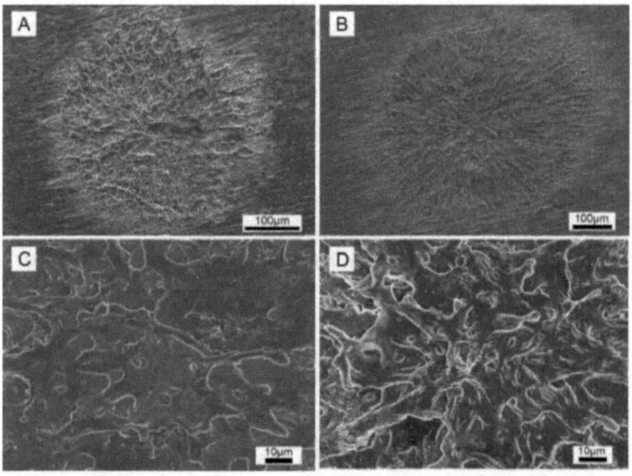

Fig. 6. Characteristic morphologies of craters in Al target produced by glass-confined ablation with different laser intensities. (a) and (c): I=1.7×10^{11} W/cm^2 ; (b) and (d): I=3.7×10^{11} W/cm^2. The top row shows overall view of the craters, and the bottom row presents detail view of the craters.

It is noteworthy that with the same laser intensity of 3.7×10^{11} W/cm^2, phase explosion only occurs in Ti target instead of Al target even though Al target has a lower critical temperature (5410 K) than Ti target (7890 K). There are three possible reasons for this phenomenon. Firstly, Al target has higher reflectivity to laser beam than Ti target such that less laser energy could reach the target surface. Secondly, owing to lower boiling temperature (2600 K) and lower ionization potentials (E_1= 5.98577 eV) of Al than those of Ti (3533 K and E_1= 6.8282 eV), denser plasma is inclined to be generated on Al target rather than on Ti target, resulting in less laser energy reaching the Al target surface through energy absorption of plasma. Thirdly, since much thicker molten surface layer is generated in Al target rather than in Ti target as aforementioned, it is more difficult for Al target than Ti target to heat all the liquid in the molten pool to near critical temperature for the occurrence of phase explosion.

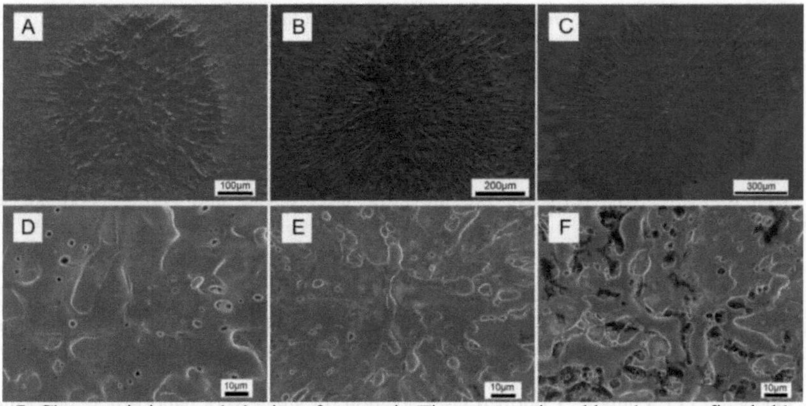

Fig. 7. Characteristic morphologies of craters in Ti target produced by glass-confined ablation with different laser intensities. (a) and (d): I=7.5×10^{10} W/cm^2 ; (b) and (e): I=1.7×10^{11} W/cm^2; (c) and (d): I=3.7×10^{11} W/cm^2. The top row shows overall view of the craters, and the bottom row presents detail view of the craters.

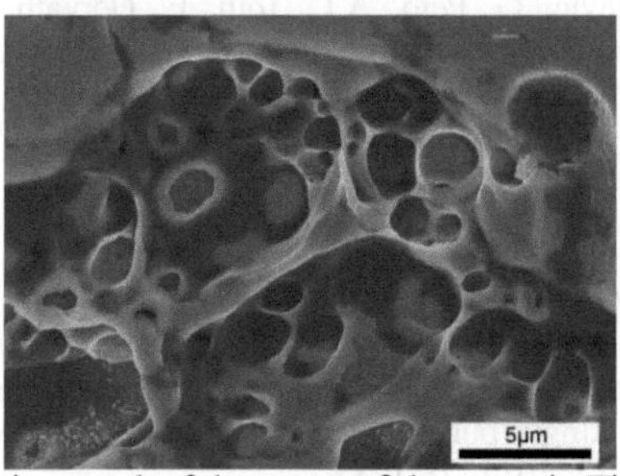

Fig. 8. Magnified SEM micrograph of the center of the crater in Ti alloy produced by glass-confined ablation with laser intensity of 3.7×10^{11} W/cm^2 (Fig. 7f).

Conclusions

In this work, two modes of laser ablation were carried out. One is the glass-semiconfined laser ablation of Al alloy using laser beam with the wavelength of 1.064 m, the other one is the glass-confined laser ablation of Al and Ti alloys using laser beam with the combined wavelengths of 532 nm and 1.064 m.

For semiconfined laser ablation, the crater profiles (depth and diameter) and the morphologies are closely related to the gap width which significantly influences the confinement degree on plasma. The smaller the gap width is, the stronger confinement on plasma is, and so is the shock wave. The diameter and depth of craters increase with the gap width decreasing.

For confined laser ablation, it is shown that the crater depth is more representative than the crater diameter as a function of laser intensity. With the same laser intensity and the same spot size, the depth of craters in Ti target is much smaller than that in Al target while the diameter of craters in Ti target is much larger than that in Al target. Besides, phase explosion occurs in Ti target instead of Al target with the laser intensity of 3.7×10^{11} W/cm^2.

Acknowledgements

The work was supported by a Tsinghua Scholarship for Overseas Graduate Studies. The authors also want to thank N.E. Kask for providing the laser device and giving good advises in experimental design.

References

1. X.Q. Wu, Z.P. Duan, H.W. Song, Y.P. Wei, X. Wang, C.G. Huang, "Shock pressure induced by glass-confined laser shock peening: Experiments, modeling and simulation," *J. Appl. Phys.*, 110 (2011) 053112 (053117 pp.)-053112 (053117 pp.).
2. T. Donnelly, J.G. Lunney, "Confined laser ablation for single-shot nanoparticle deposition of silver," *Appl. Surf. Sci.*, 282 (2013) 133-137.
3. J. Deoksuk, O. Joon Ho, L. Jong-Myoung, K. Dongsik, "Enhanced efficiency of laser shock cleaning process by geometrical confinement of laser-induced plasma," *J. Appl. Phys.*, 106 (2009) 014913 (014917 pp.)-014913 (014917 pp.).
4. S. Lugomer, B. Mihaljevic, G. Peto, A.L. Toth, E. Horvath, "Spongelike metal surface generated by laser in the semiconfined configuration," *J. Appl. Phys.*, 97 (2005).
5. R. Karimzadeh, J.Z. Anvari, N. Mansour, "Nanosecond pulsed laser ablation of silicon in liquids," *Applied Physics a-Materials Science & Processing*, 94 (2009) 949-955.
6. V. Craciun, N. Bassim, R.K. Singh, D. Craciun, J. Hermann, C. Boulmer-Leborgne, "Laser-induced explosive boiling during nanosecond laser ablation of silicon," *Appl. Surf. Sci.*, 186 (2002) 288-292.
7. Z.Y. Zheng, J. Zhang, Z.Q. Hao, X.H. Yuan, Z. Zhang, X. Lu, Z.H. Wang, Z.Y. Wei, "The characteristics of confined ablation in laser propulsion," *Chinese Physics*, 15 (2006) 580-584.
8. M.H. Mahdieh, M. Nikbakht, Z.E. Moghadam, M. Sobhani, "Crater geometry characterization of Al targets irradiated by single pulse and pulse trains of Nd:YAG laser in ambient air and water," *Appl. Surf. Sci.*, 256 (2010) 1778-1783.
9. R. Fabbro, J. Fournier, P. Ballard, D. Devaux, J. Virmont, "Physical study of laser-produced plasma in confined geometry," *J. Appl. Phys.*, 68 (1990) 775-784.

EPD Congress 2016
Edited by: Antoine Allanore, Laura Bartlett, Cong Wang, Lifeng Zhang, and Jonghyun Lee
TMS (The Minerals, Metals & Materials Society), 2016

THE COOLING ABILITY STUDY ON CO_2 AND O_2 MIXED INJECTION IN VANADIUM EXTRACTION PROCESS

Pengcheng Li[1], Yu Wang[1*], Wei-Tong Du[1], Gang Wen[1]

[1]College of Materials Science and Engineering; Chongqing University;

Chongqing 400044,China

Keywords: Vanadium extraction, O_2-N_2 Mixed injection, O_2-CO_2 Mixed injection, CO_2 cooling ability

*Corresponding author: wangyu@cqu.edu.cn

Abstract

Carbon dioxide could be utilized as a weak oxidant and a kind of coolant to oxidize elements, meanwhile, helping control the temperature during the converter vanadium extraction process. However, the optimum content of CO_2 and the cooling effect of CO_2 at low content have not been reported. In this study, experimental research based on the influence of different CO_2 contents from 0% to 25% injected to the vanadium-containing hot metal was carried out, as well as contrast experiments of O_2-N_2 mixed blowing. The results indicated that the optimum content of CO_2 was 15%. Under the optimum condition, the oxidation of [C] was the lowest and the oxidation rate of [V] was 96.9%, while the temperature was also lower than the O_2-N_2 mixed blowing. This paper provide a potential property for utilizing CO_2 during the converter vanadium extraction process.

Introduction

With an increasing attentions to the problems of global warming, many research teams are working on how to reduce the carbon dioxide emissions and taking use of it.[1-4] It is high time for iron and steel industry, which plays a big role in the emissions of CO_2, to make some changes.

Nowadays, lots of studies on applications of CO_2 during steelmaking process have been carried out. In these papers, CO_2 are used as protective gas, reaction media and stirring gas.[5-8] They found some applications will improve the quality of molten steel and rolled steel. However the application of CO_2 in the vanadium extraction process has not been reported yet.

At present, steel mills always use pure O_2 to oxidize [V] into (V_2O_3)[9] during converter steelmaking process. However, the strong reaction between O_2 and the elements in hot metal will cause the temperature to increase too highly. There will be some energy wasted and more solid coolants will be used. Those solid coolants will influent the quality of molten steel.

Therefore, it is necessary to do some studies on the cooling ability of CO_2 by using CO_2-O_2 gas as oxidizer during the converter vanadium extraction. This work was supported by National Natural Science Foundation of China (project No.51334001) and Sharing Found of Large Scale Equipment, Chongqing University (project No.201406150044).

Experimental

The vanadium-bearing metal was provided by PZH Steel, China. The chemical compositions of the vanadium-bearing metal are shown in Table 1.

Table 1. Composition of vanadium containing hot metal

composition	C	Si	Mn	V	P	S
Wt%	3.73	0.0535	0.14	0.325	0.075	0.145

Experimental Set-up

Figure 1 shows the equipment for the experiments, which includes a $MoSi_2$ electric resistance furnace and the flow rate control system. The flow rate control system consists of 4 valves, 4 flowmeters and 3 different kinds of gases: O_2, CO_2, N_2. The use of the flow rate control system is to decide which one or two gases can be blown into the furnace and control its flow rate.

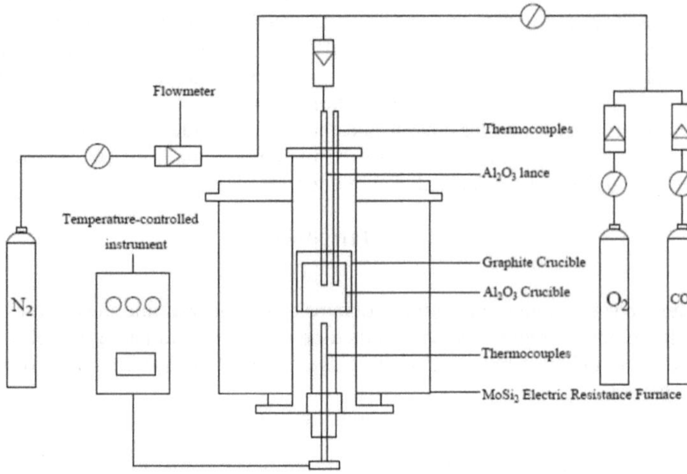

Fig. 1. Schematic drawing of experimental apparatus

Experimental Method

350 g vanadium-bearing metal was charged into a corundum crucible (inner diameter, 46 mm; height, 120 mm) which was then placed in another graphite crucible (inner diameter, 55 mm; height,135 mm). They were placed in the furnace at 1573K(1300℃) and held at 1573K for 30 min to ensure the metal melted completely. Then the blowing was began and lasted for 15min. Total flow rate of the CO_2 and O_2 mixed blowing was controlled at 0.6L/min, and with varied proportions of CO_2 as 0%, 5%, 10%, 15%, 20%, 25% in the mixed gas, the rest of which is pure oxygen.

The N_2 and O_2 mixed blowing was also done as comparison experiment. In the experiment, the proportion of N_2 is 15%.

Results and Discussion

Cooling Ability

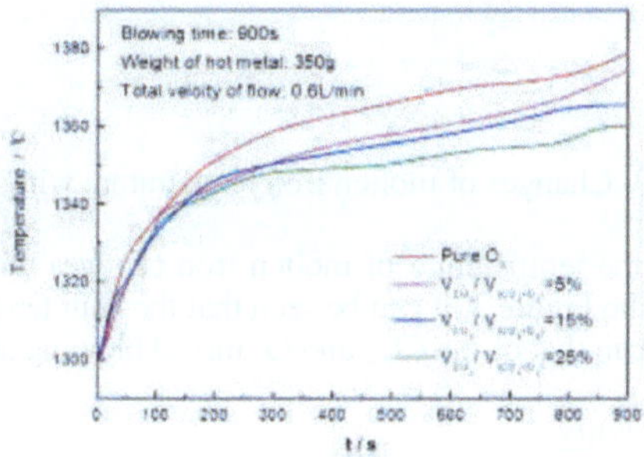

Fig 2. Changes of molten iron temperature with blowing time.

Fig. 2 shows how the temperature of molten iron changes with blowing time in different CO_2 proportion. From the figure 2, it can be seen that the bath temperature increase fast in the first 300s and tend to be constant after 300s. The bath temperature during blowing is lower with the Carbon dioxide content increasing from 5% to 25% after 300s.
The temperature of the iron bath can be influenced by CO_2 which mainly because of the following reactions.

Table 2. The thermodynamics data of interrelated chemical reactions

Chemical reaction	ΔG^0(J/mol)	ΔH^0 (kJ/kg)
$[C]+CO_{2(g)}=2CO_{(g)}$	34580-30.95T	11602.67
$[Fe]+CO_{2(g)}=(FeO)+CO_{(g)}$	11880-9.92	720
$[Si]+2CO_{2(g)}=(SiO_2)+2CO_{(g)}$	-3577967+357.27	-9299
$[Mn]+CO_{2(g)}=(MnO)+CO_{(g)}$	-261507.82+72.905T	-1512
$[Si]+O_{2(g)}=(SiO_2)$	-866510+152.30T	-29202
$[Mn]+1/2O_{2(g)}=(MnO)$	-803750+171.57T	-6594

Table 2 shows that the reactions of CO_2 and C or Fe is endothermic reactions. And the reactions of CO_2 and Si or Mn output only about 30% heat compared with that the reactions of O_2 and Si or Mn.

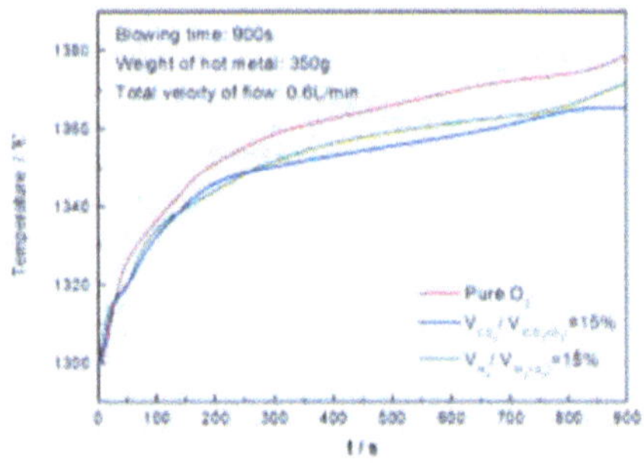

Fig 3. Changes of molten iron temperature with blowing time.

Fig. 3 shows how the temperature of molten iron changes with blowing time in different atmosphere. From the Figure 3, it can be seen that the bath temperature of N_2 and O_2 mixed blowing is higher than that of the CO_2 and O_2 mixed blowing after 300s.

Effect of Cooling Ability

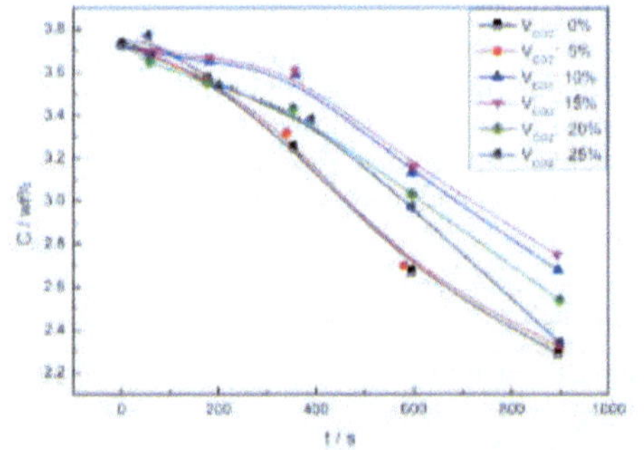

Fig 4. The [C] in molten iron changes with blowing time.

Fig. 4 shows that the [C] in the molten iron decrease continuously with blowing time. More [C] participated in the reaction as the proportion of CO_2 (volume%) increasing from 0% to 15%. However, the more proportion of CO_2 (volume%) blew, the lower [C] existed when the proportion of CO_2 (volume%) was from 15% to 25%. It is mainly because of the following reactions.

$$[C]+1/2O_{2(g)}=CO_{(g)} \qquad \Delta G^0=-136990-43.51T(J/mol) \qquad (1)$$

$$2/3[V]+1/2O_{2(g)}=1/3(V_2O_3)_{(g)} \qquad \Delta G^0=-387160+109.58T(J/mol) \qquad (2)$$

$$2/3[V]+CO_{2(g)}=1/3(V_2O_3)_{(g)}+[C] \qquad \Delta G_1^0=-250170+153.09T\ (J/mol) \qquad (3)$$

$$[C]+CO_{2(g)}=2CO_{(g)} \qquad \Delta G^0=34580-30.95T(J/mol) \qquad (4)$$

$$[C]+[O]=CO_{(g)} \qquad \Delta G^0=-17166-42.5T(J/mol) \qquad (5)$$

$$CO_{(g)}+[O]=CO_{2(g)} \qquad \Delta G^0=-131945+87T(J/mol) \qquad (6)$$

Eq.(1)-(3) shows that there is a selective oxidation temperature T_{trans}=1634K(1361℃; ΔG_1^0 =0). When bath temperature is over T_{trans}, more [C] will participate in the reaction with O_2 and the oxidation of [V] will be inhibited. However, According to other researchers' study[10], decarburization reaction of hot metal is influenced synthetically by the Eq.(4)-(6) and especially when the amount of CO_2 in the mixed gas is large. So the carbon in the molten iron decreased with CO_2 proportions increasing from 15% to 25%.

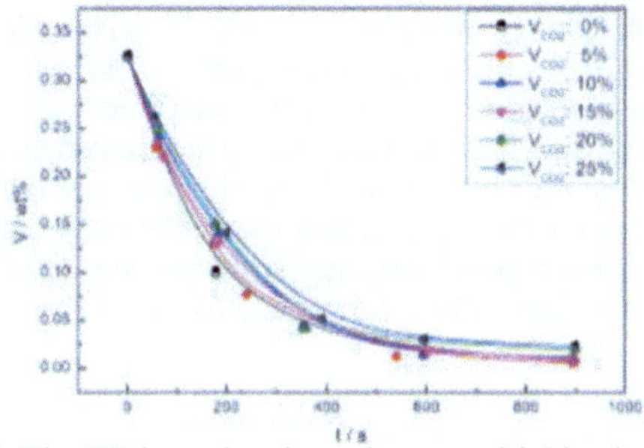

Fig 5. The [V] in molten iron changes with blowing time.

Fig. 5 shows that the [V] in the molten iron decrease continuously with blowing time and reached equilibrium state when blowing time was about 600s. It can be seen that the oxidation rate of [V] was 96.9% when the proportion of CO_2 was from 0% to 15%, while the oxidation rate of [V] was 92.1% when the proportion of CO_2 was from 20% to 25%. That is mainly because that CO_2 is a weak oxidant which will weaken the oxidation of [V]. However, all these can meet the demand of industry standard which demands the content change rate of V is more than 90%.

According to the thermodynamic analysis and the experimental results above, there is a proper proportion of CO_2 in mixed gas, which can match the objective of industry to extract more vanadium and keep more [C] in the molten steel.

The vanadium extraction process ended at 600s, as shown in Fig. 5. So the ratio of the oxidation quantity of [C] to the oxidation quantity of [V] at 600s is an important parameter to reflect the effect of vanadium extraction and carbon preservation process. The higher the ratio is, the better the vanadium extraction process is. In these experiments, the $\Delta V/\Delta C$ reached the peak value when the content of CO_2 was 15%, as shown in Fig. 6.

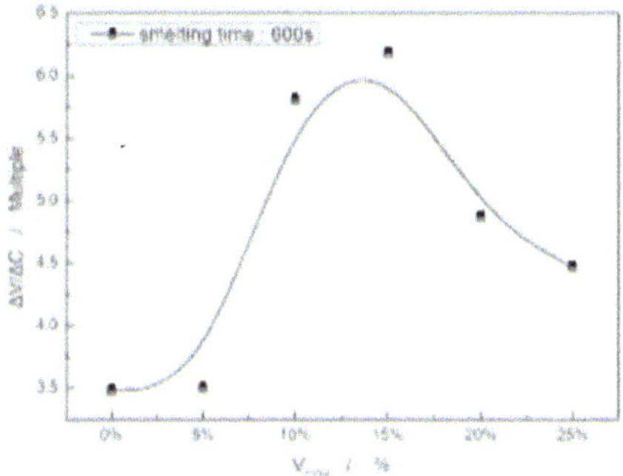

Fig.6. The Δ V/Δ C changes with the proportion of CO2(volume%) at 600s

Conclusions

Based on the theoretical analysis and lab-scale experimental study above, the following conclusions can be drawn.

[1] The test of temperature control in CO_2 and O_2 mixed blowing vanadium extraction process shows that the bath temperature (after 300s) has continued to decrease as increasing the proportion of Carbon dioxide. Compared with N_2 and O_2 mixed blowing, CO_2 and O_2 mixed blowing has a better effect on the temperature controlling.

[2] The cooling ability of CO_2 has remarkable effect on vanadium extraction process. With the proper proportion of CO_2, more vanadium will be extracted and more [C] will be kept in molten steel. In these experiments, the proper proportion of CO_2 is 15%, under which condition the oxidation rate of [V] is 96.9% and the $\Delta V/\Delta C$ reach the peak value.

References

[1] Grimston M. C. et al., "The European and Global Potential of Carbon Dioxide Sequestration in Tackling Climate Change," *Climate policy*, 2 (2001), 155-171.

[2] Wang K. et al., "Scenario Analysis on CO2 Emissions Reduction Potential in China's Iron and Steel Industry," *Energy Policy*, 35(4) (2006), 2320–2335.

[3] Hashimoto K. et al., "Materials for Global Carbon Dioxide Recycling," *Corrosion Science*, 2(2002), 371-386.

[4] Worrell, Ernst, Lynn Price, and Nathan Martin. "Energy Efficiency and Carbon Dioxide Emissions Reduction Opportunities in the US Iron and Steel Sector," *Energy*, 26(5) (2001): 513-536.

[5] Bruce T, Weisang F, Allibert M, et al. "Effects of CO2 Stirring in a Ladle, " (Paper presented at the Electric Furnace Conference Proceeding, Chicago, 1987).

[6] Hiryukt Katayama, Yasuhisa Abe. " Production of Low Nitrogen Steels, " (Paper presented at the International Symposium on the Physical Chemistry of Iron and Steelmaking(CIM), Toronto, 1982)

[7] Hara R D O,Spence A G R, Eissen wasser J D. "Carbon Dioxide Shrouding and Purging at IPSCO's Melt Shop, " *Iron and Steelmaker*, 1986,13(3):24.

[8] Yi C, Zhu R, Chen B Y, et al. "Experimental research on reducing the dust of BOF in CO2 and O2 mixed blowing steelmaking process, "*ISIJ Int*, 2009, 49(11): 1694

[9] Moskalyk, R.R., and A.M. Alfantazi. "Processing of Vanadium: A Review, " *Minerals Engineering*, 16(9) (2003), 793-805.

[10] Nomura H, Mori K. "Kinetics of Decarburization of Liquid Iron with High Concentration of Carbon," *Tetsu-to-hagané*, 57(9) (1971), 1468.

EPD Congress 2016
Edited by: Antoine Allanore, Laura Bartlett, Cong Wang, Lifeng Zhang, and Jonghyun Lee
TMS (The Minerals, Metals & Materials Society), 2016

Effect of MnO on Sintering and Microstructure of Al_2O_3-MgO-CaO Refractories

Xue-liang Yin[1,2], Lei Liu[1], Xiang Shen[1], Mei-le He[1], Lei Xu[1], Nan Wang[1], Min CHEN[1]*

(1 School of Materials and Metallurgy, Northeastern University, Shenyang 110819, China; 2 Liao Ning Institute of Science and Technology, Benxi 117000, China)

Keywords: Al_2O_3-MgO-CaO, refractory, MnO addition, sintering, microstructure

Abstract

In the current study, Al_2O_3-MgO-CaO refractories were prepared at 1400-1600°Cby the addition of MnO micro-powders, and the effect of MnO addition on densification behavior of the refractory was discussed. The results showed that the doped MnO dissolved to $MgAl_2O_4$ phase, and promoted the growth of $MgAl_2O_4$ grains by the formation of $MgAl_2O_4$ solid solution. As a result, the dense microstructure was obtained, with the apparent porosity decreased from 19.2% to 5.4% and the bulk density increased from 2.78g/cm^3 to 3.15g/cm^3 after firing at 1600℃ for 2h by the addition of 4% MnO. In addition, a texture microstructure was observed, which is considered to be favorable to improve mechanical properties and the service life of Al_2O_3-MgO-CaO system refractories.

Introduction

In the Al_2O_3-MgO-CaO ternary phase diagram, the appearance temperature of liquid phase is higher than 1730℃, making use of the thermodynamic information provided by the Al_2O_3-MgO-CaO system, high refractoriness refractories can be designed. Especially when they are located in the subsystem Al_2O_3-$MgAl_2O_4$-$CaAl_4O_7$ area, the appearance temperature of liquid phase is up to 1850 ± 10℃[1,2], and the true density of CaO·$2Al_2O_3$ is smaller in this system[3,4]. Therefore, the high temperature performance can be ensured by adding a small quantity cheap lime (CaO) to replace part of the Al_2O_3 raw materials into the Al_2O_3-MgO system, the lightweight refractory of the Al_2O_3-MgO-CaO system can be prepared, while the lightweight refractory necessarily means energy conservation and emissions reduction [5]. Effects of CaO content on sintering and lightweight of Al_2O_3-MgO-CaO refractories was studied in our previous work[6]. In the synthesis process of the Al_2O_3-MgO-CaO system refractories, sintering and densification is suppressed with the volumetric expansion of multiple reaction between each component, it is difficult to obtain dense refractory. Particularly, spinel has the highest melting point in the phase composition of the Al_2O_3-MgO-CaO system refractories, it is also the most difficult to sinter by using single step sintering method, and general requirements sintering temperature is not lower than 1700℃ even by two-stage sintering method, which is the main reason for restricting its development[7,8].

Based on these aspects described above, in the current study, Al_2O_3-MgO-CaO refractories was prepared by two-stage sintering method, and the effect of MnO addition on sintering and microstructure of Al_2O_3-MgO-CaO refractories were investigated.

Experiment Procedures

The experimental batch compositions are shown in Table 1. The starting raw materials were 82% Al_2O_3 (particle size 8-12 m, 99% purity; Chinalco, Beijing, China), 10% MgO (particle size 2-3 m, 99% purity; Kermel Chemical Reagent, Tianjing, China), $CaCO_3$ (particle size 3-5 m, 99.5% purity; Kishida Chemical, Osaka, Japan) calculated on the ratio of 8% CaO.

Table 1. Batch composition of the samples/ (mass%)

Batchs	$CaCO_3$/%	Conversion CaO/%	MgO/%	Al_2O_3/%	MnO/%
0#	14.29	8	10	82	0
1#	14.29	8	10	82	2
2#	14.29	8	10	82	4
3#	14.29	8	10	82	6

The mixtures according to the preparation of 0 batch were ground in a laboratory-scale attrition milling in isopropanol media for 4h, to obtain homogeneous and highly energetic powder mixtures, and the mixed powder were dried at 120℃, then were isostatically pressed at 25MPa to bars (20mm×20mm×10mm). Pressed shapes were sintered at 1200℃ with a heating rate of 5℃ per min and cooling with furnace after a soaking period of 2h at the peak temperatures, then to grind the first stage sintered samples in a pot mill for 1h, and MnO powder(particle size 3-5 m, 99.5% purity; Xiya Reagent, Shandong, China) was added to the above composite, with external addition of 0, 2, 4 and 6% (shown in Table 1) respectively. Subsequently attrition milled for 4h, mixed powders were then isostatically pressed at 40MPa to briquettes (*D*20mm×6mm). These prepared samples were sintered secondly in a high temperature electric furnace at 1400, 1500 and 1600℃ in air atmosphere with a heating rate of 5℃ per min and cooling with furnace after a soaking period of 2h at the peak temperatures.

Briquettes were used to characterize the densification, phase compositions, microstructures. The densification was characterized by bulk density and apparent porosity measured in kerosene using Archimedes principle. The phase compositions were examined by X-ray diffraction method(XRD; X'pert PRO, PANalytical, Netherlands) using Cu $K\alpha 1$ radiation (λ =1.5406 Å) with a step of 0.02°(2θ) and a scanning rate of 2°/min from range of 10° to 90°. The microstructures were analyzed by scanning electron microscopy (SEM; EVO-18, ZEISS, Germany).

Results and Discussions

XRD patterns of the first stage samples sintered at 1200°C for 2h is shown in Fig.1. It is observed that the presence of reaction products CA_2, MA and unreacted periclase, corundum phases. It may be explained that the solid reaction is not thoroughly during the sintering of the first stage samples.

XRD patterns of the second stage sintered samples for both with 2%, 4%, 6% and without additives at 1600℃ for 2h (Fig.2a) shows the phase of the samples are MA, CA_2 and CA_6, no diffraction lines for MnO were detected for the case of 2%, 4% and 6% MnO added. Further analysis shows that only the diffraction peak of MA took place shift among the three kinds of peak position. This implies that the added MnO might have formed $MgAl_2O_4$-MnO solid solution. By reason of $R_{Mn}{}^{2+}$(0.067nm)>$R_{Mg}{}^{2+}$(0.049nm), Mn^{2+} having similar characteristics can substitute the Mg^{2+} in spinel, Mn^{2+} will causes the lattice distortion of the Mg^{2+} crystal cell, which is considered , it can result in fast diffusion within the single grains during the sintering process, which is favorable to improve the reaction speed. Meanwhile, the lattice parameters and the interplanar spacing of Mg^{2+} crystal cell will increase during the process of the lattice distortion. In combination with bragg equation $\lambda = 2d\sin\theta$, Mn^{2+} will prompt the diffraction peak of MA moving to a lower angle. Fig.2b is the slow-scan patterns of (400) of MA solid solution, reproducing an enlarged section of Fig.2a, it shows that the peak of (400) shifted to a lower angle with increased amount of MnO to 6%[9]. This result precisely proved the above analysis and identified the formation of the $MgAl_2O_4$-MnO solid solution.

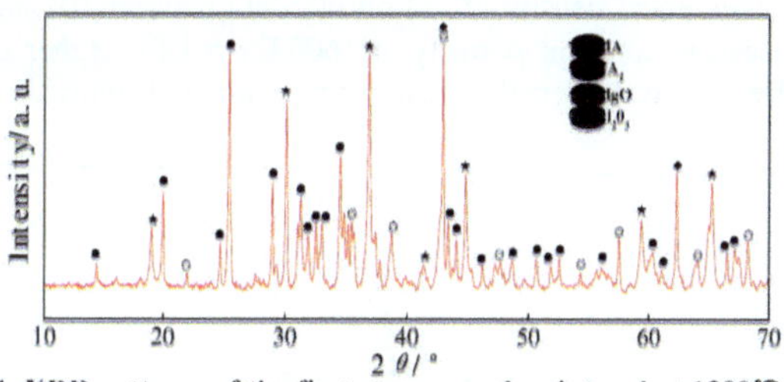

Fig. 1. XRD patterns of the first stage samples sintered at 1200℃ for 2 h.

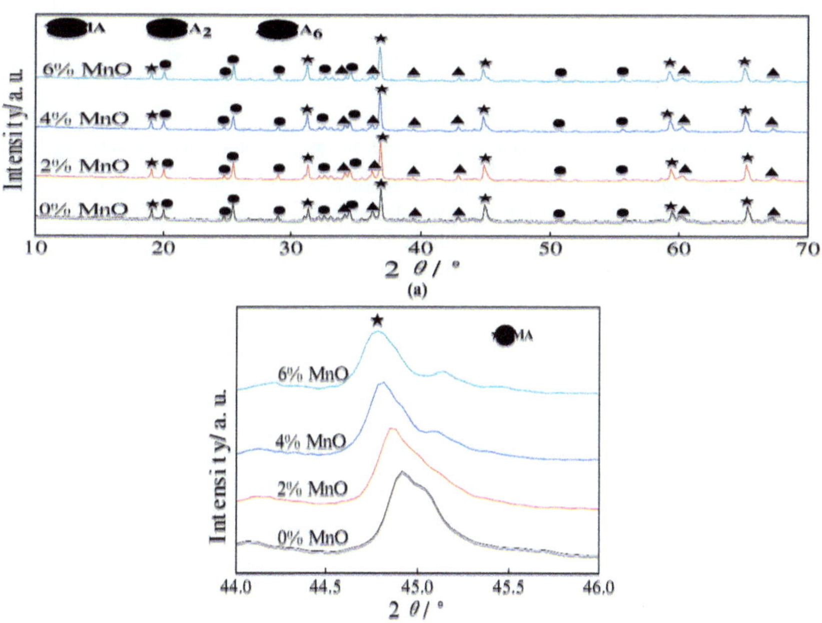

Fig. 2. XRD patterns of different MnO content second stage samples sintered at 1600°C for 2h.(a) normal XRD patterns,(b) XRD pattern of (400) of MA crystal.

Fig.3 shows an increase in the bulk density and a reduction in the apparent porosity with the increase in the sintering temperature and additive amount of the two stage sintered samples. Sintering below 1500℃ is not sufficient to obtain high densification in samples with and without additive. Composition without additive results in a very poor density and a very high apparent porosity on sintering even at 1600℃. Again MnO too improves the density and reduces the apparent porosity sharply, which was associated with the grain growth and pore coalescence. For the batch sintered at 1600℃ the beneficial effect of additive of 4% and 6% MnO was most prominent. The apparent porosity decreased from 19.2% to 5.4% and the bulk density increased from 2.78g/cm^3 to 3.15g/cm^3 after firing at 1600℃ for 2h by addition of 4% MnO, it has reached more than 90% of the theoretical density. Addition of MnO between 0% and 4% enhances the densification and reduces the apparent porosity at 1600℃, but for higher additive amount the scope of further increase in sintered density and apparent porosity is limited.

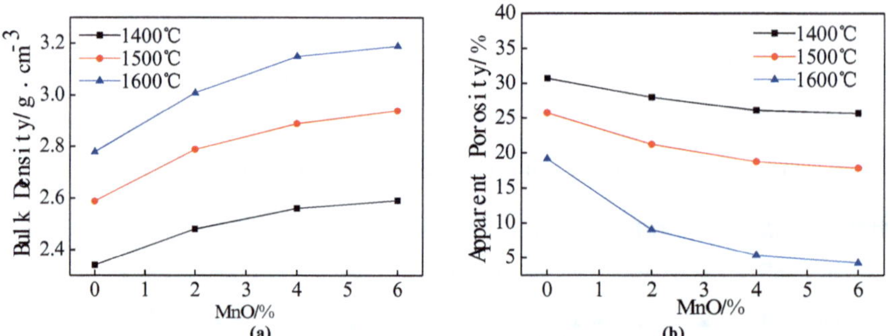

Fig. 3. Variation of bulk density and apparent porosity of the samples after two-stage sintering.

The typical back-scattered electron (BSE) images of microstructures on the polished surfaces of the with and without MnO content second stage samples at 1600℃ are given in Fig.4a, it shows that the microstructure of sintered samples without MnO content samples is not uniform, a porous structure was observed, with limited grain growth. However, the intergranular pores amount of the 4% MnO content samples sintered at 1600℃(Fig.4b) shows an obviously reduce, and the grain size shows an obviously increase in the MA of octahedrons structure and quadruple prisms structure of CA_2 with the increase of the MnO additive, but it shows a obviously reduce in the plates structure of CA_6 grain size. Analysis the above reason is that the doped MnO dissolved to $MgAl_2O_4$ phase, and the sintering activity of $MgAl_2O_4$ phase was obviously improved by the formation of $MgAl_2O_4$-MnO solid solution. So it mainly promote the spinel grain growth and occupy a certain space, therefore it limit the growth space of CA_6, thereby it inhibit the growth of CA_6 grain[10]. On the contrary, it promote the growth of the CA_2 grain because of it experience the formation and growth ($CaO \cdot 2Al_2O_3 + 4Al_2O_3 = CaO \cdot 6Al_2O_3$) during the CA_6 grain formation process within the system. In addition, a texture microstructure of the crystal phase, which includes MA, CA_2 and CA_6, was observed in the typical back-scattered electron (BSE) images of microstructures with 4% MnO content samples sintered at 1600℃ for 2h. Which is considered to be favorable to improve the mechanical properties and the service life of the Al_2O_3-MgO-CaO system refractories[11,12].

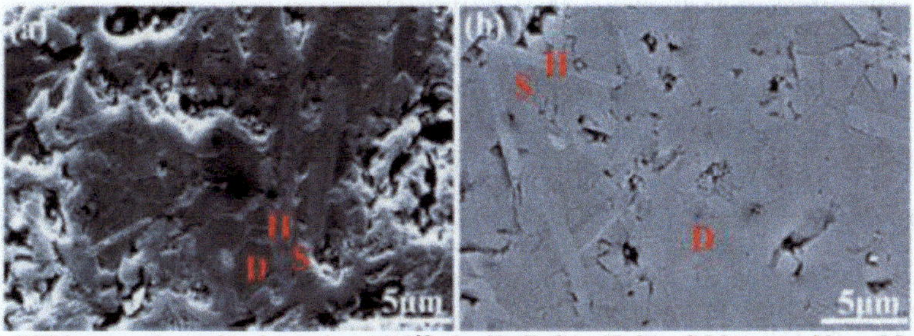

Fig. 4. BSE images of samples sintered at 1600℃ for 2h.(D:CA_2, H:CA_6, S:MA):(a) without MnO. (b) 4% MnO

Conclusions

Al_2O_3-MgO-CaO refractories was prepared by two-stage sintering method, and the effect of MnO addition on sintering and microstructure were investigated. Based on the above results, the following conclusions have been drawn:

1) The phase compositions of Al_2O_3-MgO-CaO refractories after firing at 1600℃ for 2h are MA, CA_2 and CA_6.
2) The added MnO causes the lattice distortion of MA crystal, which is considered , it can result in fast diffusion within the single grains during the sintering process, which is favorable to improve the reaction speed.
3) The added MnO dissolved to $MgAl_2O_4$ phase, and promoted the growth of $MgAl_2O_4$ grains by the formation of $MgAl_2O_4$ solid solution. As a result, the dense microstructure was obtained, with the apparent porosity decreased from 19.2% to 5.4% and the bulk density increased from 2.78g/cm^3 to 3.15g/cm^3 after firing at 1600℃ for 2h by addition of 4% MnO.

Acknowledgements

The authors gratefully acknowledge the National Natural Science Foundation of China (Nos. 51174049, 51174052, 51374057 and 51374062) to support the current research.

References

1. A. H. De Aza, J. E. Iglesias, P. Pena, et al., "Ternary system Al_2O_3-MgO-CaO: part II, phase relationships in the subsystem Al_2O_3-$MgAl_2O_4$-$CaAl_4O_7$," Journal of the American Ceramic Society, 83(4) (2000), 919-927.
2. T. Durán, S. Serena, P. Pena, et al., "Experimental establishment of the $CaAl_2O_4$-MgO and $CaAl_4O_7$-MgO isoplethal sections within the Al_2O_3-MgO-CaO ternary system," Journal of the American Ceramic Society, 91(2) (2008), 535-543.
3. A. Ghosh, S. K. Das, J. R. Biswas, et al., "The effect of ZnO addition on the densification and properties of magnesium aluminate spinel," Ceramics International, 26(6) (2000), 605-608.
4. A. Altay, C. B. Carter, P. Rulis, et al., "Characterizing CA_2 and CA_6 using elnes," Journal of Solid State Chemistry, 183(8) (2010), 1776-1784.

5. W. Yan, N. Li and B. Q. Han, "High-strength lightweight spinel refractories," American Ceramic Society Bulletin, 84(4) (2005), 9201-9203.
6. L. Xu, M. Chen,W. J. Huang, et al., "Effects of CaO content on sintering and lightweight of Al_2O_3-MgO-CaO refractories," Materials Research Innovations, 19(s5) (2015), 212-217.
7. S. Sinhamahapatra, K. Dana, A. Ghosh, et al., "Dynamic thermal study to rationalize the role of titania in reaction sintering of magnesia-alumina system," Ceramics International, 41(1) (2015), 1073-1078.
8. H. S. Tripathi, S. Singla, and A. Ghosh, "Synthesis and densification behaviour of magnesium aluminate spinel: Effect of Dy_2O_3," Ceramics International, 35(6) (2009), 2541-2544.
9. M. Chen, A. J. Jin, N. Wang, et al., "Synthesis of hydration-resistant CaO refractory by addition of MgO," Developments in Chemical Engineering and Mineral Processing, 14(3/4) (2006), 409-416.
10. C. Domínguez, J. Chevalier, R. Torrecillas, et al., "Microstructure development in calcium hexaluminate," Journal of the European Ceramic Society, 21(3) (2001), 381-387.
11. L. A. Díaz, R. Torrecillas, A. H. De Aza, et al., "Effect of spinel content on slag attack resistance of high alumina refractory castables," Journal of the European Ceramic Society, 27(16) (2007), 4623-4631.
12. A. P. Luz, M. A. L. Braulio, A. G Tomba Martinez, et al., "Slag attack evaluation of in situ spinel-containing refractory castables via experimental tests and thermodynamic simulations," Ceramics International, 38(2) (2012), 1497-1505.

EPD Congress 2016
Edited by: Antoine Allanore, Laura Bartlett, Cong Wang, Lifeng Zhang, and Jonghyun Lee
TMS (The Minerals, Metals & Materials Society), 2016

DETERMINATION OF TOTAL IRON CONTENT IN IRON ORE AND DRI: TITRIMETRIC METHOD VERSUS ICP-OES ANALYSIS

Yousef Mohassab, Mohamed Elzohiery, Feng Chen, and Hong Yong Sohn

Department of Metallurgical Engineering, University of Utah, Salt Lake City, Utah 84112, USA

Keywords: ICP, Titration, Flash ironmaking, chemical analysis

Abstract

The determination of reduction degree in a DR process is sensitive to the total iron in the ore and DRI. An accurate and high throughput analysis method for total iron has been developed. Titration of the solution after tin(II) chloride reduction of ferric ion is a widely used method for iron analysis. However, it is a multistep method that requires many chemical reagents and much time. In this work, an ICP-OES analysis method with higher or equivalent accuracy compared with the titrimetric method was developed. This method has much higher throughput and demands fewer chemical reagents compared with the titrimetric method. In this paper, a comparison of the two methods is presented.

Introduction

As part of developing a novel flash ironmaking process at the University of Utah [1-10], analysis of a massive number of iron samples is carried out on a regular basis to determine the total iron content. One of the most widely used methods for iron content analysis is the titrimetric method, especially the method involving the reduction of ferric ion by tin(II) chloride (International Standard ISO 2597-1). This method is a multistep procedure requiring the preparation of an overwhelming number of chemical reagents, consuming much time, and challenging in detection of the end point. Regardless of the efforts to develop simpler titrimetric methods for the determination of total iron, titration-based methods still suffer from low throughput [11]. In this laboratory, a method that uses inductively coupled plasma-optical emission spectroscopy (ICP-OES) was developed to determine the total iron content in iron ores and reduced samples. Moreover, HF required in the titrimetric analysis of samples containing a significant content of silica reacts severely with the glassware. Replacing glassware with HF-resistant materials such as plastics is very challenging due to the difficulty in detecting the endpoint due to their opacity. A method that circumvents this problem is to convert silica into soluble salts, which is a very time consuming procedure requiring additional chemicals reagents and high temperature treatment of the samples. Bypassing the use of HF in the titrimetric method introduces significant errors in total iron determination as will be shown subsequently. In this work, an ICP-based method developed in this laboratory was compared with the ISO standard titrimetric method involving tin(II) chloride reduction. A brief description of both methods is presented and the results of the methods are compared.

Compared with the titrimetric method (ISO 2597-1), the ICP-based method requires a significantly fewer steps and chemical reagents, obviates the difficult and error-prone visual detection of the end points, allows high throughput rates, and provides equivalent or better accuracy and precision compared with the titrimetric method.

Experimental Work

Table I shows the various iron-containing samples and their total iron content used for analysis in this study. Certified reference material (CRM #690) of a Canadian iron ore concentrate was obtained from the National Institute of Standards and Technology (NIST) in Boulder, Colorado, USA. Hematite (99.945%) and magnetite (99.99%) reference materials supplied by Alfa Aesar (Ward Hill, MA, USA) and Sigma Aldrich (St Louis, MO, USA). In addition, hematite concentrate ore from the Yuanjiacun Range, Shanxi Province, China and the flash reduced samples produced from it in a high temperature drop tube reactor [9, 10]. Flash reduced samples of magnetite concentrate from the Mesabi Range (U.S.) were also analyzed by both methods. The particle sizes of all samples were < 50 m. All the samples were well mixed using a mechanical mixer for about 10 minutes in order to ensure the homogeneity of the samples before sampling for analysis. The accuracy of the analytical balance was ± 0.5 mg. All samples were analyzed at least three times with both methods.

Table I. Chemical analysis (in mass fraction) of the samples used in the analysis.

Samples	Code	Total Iron	SiO_2	Al_2O_3	P_2O_5
Certified Reference Sample	CRM	0.6685	0.0371	0.018	0.025
Pure Fe_2O_3 Reference Sample	RM 1	0.699	-	-	-
Pure Fe_3O_4 Reference Sample	RM 2	0.724	-	-	-
Fe_2O_3 Chinese Ore	H	0.657	<0.05	<0.008	<0.0007
Flash Reduced Magnetite with High Reduction Degree	FRMH	0.90	<0.03	<0.01	
Flash Reduced Magnetite with Low Reduction Degree	FRML	0.75	<0.03	<0.01	
Flash Reduced Hematite with High Reduction Degree	FRHH	0.90	<0.05	<0.008	<0.0007
Flash Reduced Hematite with Low Reduction Degree	FRHL1	0.72	<0.05	<0.008	<0.0007
Flash Reduced Hematite with Low Reduction Degree	FRHL2	0.67	<0.05	<0.008	<0.0007

Sample Analysis by Titrimetric Method after Tin(II) Chloride Reduction (ISO 2597-1)

We summarize here major steps of the titrimetric method to facilitate the comparison with the ICP method developed in this work. In the titrimetric method, iron oxide samples were digested in hydrochloric acid and reduced to Fe^{2+} by $SnCl_2$. Then, Fe^{2+} is titrated with a potassium dichromate solution of known concentration as shown by Eqs. [1] and [2]. When all Fe^{2+} is consumed by potassium dichromate, violet color indicates the endpoint in the presence of sodium diphenylaminesulfonate (indicator):

$$2Fe^{3+} + Sn^{2+} = 2Fe^{2+} + Sn^{4+} \quad [1]$$

$$6Fe^{2+} + Cr_2O_7^{2-} + 14H^{+} = 6Fe^{3+} + 2Cr^{3+} + 7H_2O \quad [2]$$

Chemical solutions used in the titration process were prepared according to the ISO standard method (International Standard ISO 2597-1), which are as follows

1. Hydrochloric acid, HCl, (ρ = 1.19 g/mL at 25 °C): Dilute 1:10 with deionized water (DI).
2. Sulfuric + phosphoric acid, H_2SO_4+H_3PO_4, mixture (H_2SO_4, ρ=1.84 g/mL, 150 mL; H_3PO_4, ρ=1.7 g/mL, 150 mL): 150 mL of H_2SO_4 are cautiously poured into about 300 mL of water while stirring, cooled in a water bath, then 150 mL of H_3PO_4 are added and diluted to 1 L with water.
3. Tin(II) chloride, $SnCl_2$, 100g/L solution: 100 g of $SnCl_2$·H_2O are dissolved in 200 mL of HCl (ρ=1.16~1.19g/mL) by heating the solution in a water bath. The solution is cooled and diluted to 1 L with water. Then the solution is stored in a brown glass bottle.
4. Mercury (II) chloride, $HgCl_2$, 50g/L solution: 50 g of $HgCl_2$ are dissolved in 1 L deionized water.
5. Iron standard solution, 0.1 mol/L: 5.58 g of pure iron are weighed into a conical flask and a small filter funnel is placed in the neck. 75 mL of HCl (ρ=1.16~1.19g/mL, diluted 1:1) are added and heated until the iron is dissolved. The solution is cooled and oxidized with 5 mL of H_2O_2 (30% by volume), then heated to a boil, transferred to a 1000 mL volumetric flask, and diluted to volume with water.
6. Potassium dichromate, $K_2Cr_2O_7$, 0.01667 mol/L solution: $K_2Cr_2O_7$ powder is dried in an air bath at 140~150 °C for 2 h and cooled to room temperature in a desiccator. 4.904 g of this dried $K_2Cr_2O_7$ is dissolved in water and then the solution is diluted to exactly 1000 mL. Temperature is recorded (T_1) at which this dilution is made.
7. Sodium diphenylaminesulfonate, $C_6H_5NHC_6H_4SO_3Na$, 2 g/L; It is stored in a brown glass bottle.

The samples were carefully dried to avoid further iron oxidation. In the ISO standard, it was specified that the mass of the samples should be within the range 0.20-0.25 g. In this work, a sample mass in the range of 0.10-0.15 g was determined to be the optimum level. The samples were digested in 30 mL HCl (Reagent 1) in a conical flask covered with a watch glass on a heating plate. The temperature was kept at ~80 °C for about 1 h until the ore was substantially digested. The glass cover was then washed by warm DI water to ensure all the condensate from the decomposed solution vapor was recovered back into the conical flask. Then the solution was diluted to 50 mL with warm DI water and heated to just below the boiling. The solution color now was clear yellow with some precipitated undigested silicates.
Hot HCl (Reagent 2) was used to wash the glass cover and the inside wall of conical flask then $SnCl_2$ solution (Reagent 4) was immediately added drop by drop into the decomposed solution until the yellow color disappeared. An additional drop of $SnCl_2$ solution (Reagent 4) was added to make sure that all the Fe^{3+} iron was reduced to Fe^{2+}.

After the solution was cooled in a water bath, 8 mL $HgCl_2$ (Reagent 5) was added to oxidize the excess $SnCl_2$ solution (Reagent 4) and mixed gently for 5 minutes. Then 30 mL H_2SO_4+H_3PO_4 mixtures (Reagent 3) was added to the solution after it was diluted to 150 mL with cold DI water. Five drops of $C_6H_5NHC_6H_4SO_3Na$ (Reagent 8) was added to the solution as an indicator and then titration with $K_2Cr_2O_7$ solution (Reagent 7) was conducted. The endpoint was obtained when the color of the solution changed to dark green and then to a violet color with one more drop of the titrant. The temperature at which the $K_2Cr_2O_7$ solution was used (T_2) and its volume used (V_1) were recorded to be used in subsequent calculations.
All the analyses were carried out at least three times at different times of the day on different days on randomly selected samples to minimize operator's systematic errors associated with the timing of the analysis or the level of skills.
In additions to the samples listed in Table I, one blank test was carried with each day's analysis under the same conditions. As $C_6H_5NHC_6H_4SO_3Na$ (Reagent 8) does not react with $K_2Cr_2O_7$ solution (Reagent 7) in the absence of iron, 1 mL iron standard solution (Reagent 6) was added in the blank solution with a 3-mL disposable pipette immediately before the addition of $SnCl_2$ solution (Reagent 4), which would promote indicator response in the blank solution and thus allow a suitable correction for the blank.
The blank test value was determined using the same amounts of all reagents and following all the steps of the procedure. The volume of $K_2Cr_2O_7$ solution (Reagent 7) used in the blank test was recorded as V_0. A volume of 1 mL of standard iron solution (Reagent 6) is equivalent to 1 mL of $K_2Cr_2O_7$ solution (Reagent 7) according to their concentration as prepared earlier, Eq. [2]. Thus, the blank test value of this titration (V_2) is calculated using the relationship $V_2 = V_0 - 1.00$.
At the end, the total iron content was calculated using the following relationship:

$$wFe = \frac{(V_1 - V_2)[1-(T_2 - T_1)\times 0.0002]}{m} \times 0.0055847 \times 100 \times K \qquad [3]$$

$$K = \frac{100}{100 - A} \qquad [4]$$

where
wFe: mass fraction of Fe in sample,
T_1: temperature when $K_2Cr_2O_7$ was prepared, °C,
T_2: temperature when $K_2Cr_2O_7$ was used, °C,
V_1: volume of $K_2Cr_2O_7$ consumed in sample titration, mL,
V_2: volume of $K_2Cr_2O_7$ consumed in blank titration, mL,
m: mass of sample, g,
A: moisture content, as a percentage by mass, determined in accordance with ISO 2596, and
K: conversion factor which is 1.00 for pre-dried test samples according to Eq. [4].

The above procedure is validated using analytical grade dry iron powder to an accuracy of 0.02 %.

Determination of Iron Fraction Using ICP-OES

Samples were prepared by accurately weighing around 0.2 g of each material in polypropylene disposable centrifuge tube. The weight of the empty tube and the sample were recorded. Under the fume hood, 15 mL of HCl (ρ = 1.19 g/mL) were added into each tube using digital micro pipette to digest the sample. For high silica content samples, 2 mL of HF (ρ = 1.15 g/mL) were added. The tubes were left open for about 20 minutes under the fume hood to release the produced gases in order not to pressurize the tubes when closed. Tubes were then closed and placed in a hot water bath until a solution with a clear yellowish solution with precipitates was observed. The solutions were left to cool down and then well mixed by a mechanical mixer for 10 minutes to ensure solution homogeneity. The solutions were then diluted to about 150 times using HCl (5% by volume). The HCl solution was prepared using HCl acid (ρ = 1.19 g/mL) and DI water which were stirred for at least 2 hours before use.

Using 3-mL disposable transfer pipettes, around 0.2 g of the digested sample solution were added to a new polypropylene disposable centrifuge tube and then diluted by the 5% HCl solution till the overall solution weight was approximately 30. g. The weight of the transferred solution and the final diluted solution were recorded to calculate the actual dilution factor.

To calibrate the ICP machine, four calibration solutions were prepared with concentrations 0, 50, 80 and 100 ppm Fe in 5% HCl solution. In order to prepare these solutions, 500 ppm Fe standard solution in 5% HCl solution supplied by Inorganic Ventures (Christiansburg, VA, USA) was further diluted using the 5% HCl. All the diluted samples were well mixed by mechanical mixer for at least 10 minutes. Using the calibration solutions, the machine was calibrated and the calibration curve (R^2) value for the calibration linear curves should be at least 0.99999.

Drift sample and reference sample were analyzed every 5 samples to detect drift or error for the human operator or the machine.

The result obtained by the machine is the concentration of iron in the diluted solution in ppm. In order to calculate the iron fraction, the following equations were used:

$$wFe = C \times 10^6 \times m_{dil} \times F_{dil} \quad [5]$$

$$F_{dil} = \frac{\text{weight of diluted solution}}{\text{weight of the digested solution sample}} \quad [6]$$

$$\omega_{Fe} = \frac{wFe}{m_s} \quad [7]$$

where

wFe: weight of iron in the diluted solution,

C: concentration of iron in the diluted solution in ppm,

m_{dil}: weight of the diluted solution,

F_{dil}: dilution factor,

m_s: weight of the powder sample, and

ω_{Fe}: iron fraction in the sample.

Both analysis methods yield the total iron fraction of the samples. The calculated iron fractions of the samples were compared to the actual iron fraction. The samples standard deviation *SD* and relative standard deviation *RSD* (%) were calculated according to the following equations:

$$SD = \sqrt{\frac{\Sigma\,(X - \bar{X})^2}{N - 1}} \quad [8]$$

$$\bar{X} = \frac{\Sigma\,X}{N} \quad [9]$$

$$RSD\% = \frac{SD}{\bar{X}}\,!\,100 \quad [10]$$

where

SD: standard deviation,
X: iron fraction,
$\bar{X}$: arithmetic mean of X,
N: total number of repeated runs, and
$RSD\%$: percent relative standard deviation.

Results

Tables II and III shows the results of both analysis methods with the calculated *SD* and %*RSD*. Comparison of the %*RSD* values from the two methods indicates that the methods have similar precision for samples with no or extremely low silica contents, as shown in Figure 1(a). For samples with high silica contents (> 5 mass %), the ICP method gave results with much greater precision than the titrimetric method, as shown in Figure 1(b). The two methods were found to be of similar accuracy, as Figure 2 shows. It is worth noting that a skilled operator can analyze at most 6-10 samples in an 8-hour period using the titrimetric method, whereas 30 – 50 samples can be analyzed using the ICP method in the same amount of time.

Table II. Analysis results of the titration method

Sample	Reference Iron Fraction	Analysis 1	2	3	4	5	6	Mean	SD	RSD (%)	Diff. (%)
CRM	0.6685	0.6693	0.6690	0.6727				0.670	0.002	0.303	0.273
RM1	0.7235	0.7272	0.7276	0.7276				0.727	0.000	0.031	0.541
RM2	0.6991	0.7012	0.6902	0.7075	0.6926	0.6908	0.7033	0.698	0.007	1.052	0.212
FRMH	N/A	0.9072	0.9047	0.8977				0.903	0.005	0.543	
FRML	N/A	0.7481	0.7488	0.7558				0.751	0.004	0.565	
H	N/A	0.6592	0.6453	0.6504				0.652	0.007	1.079	
FRHL1	N/A	0.7243	0.7067	0.7151				0.715	0.009	1.233	
FRHH	N/A	0.6415	0.5904	0.6054				0.612	0.026	4.285	

Table III. Analysis results of the ICP method

Sample	Expected Iron Fraction	Analysis 1	2	3	4	5	6	7	Mean	SD	RSD (%)	Diff. (%)
CRM	0.6685	0.6719	0.6751	0.6693	0.6752	0.675	0.6716		0.6730	0.0025	0.3646	0.676
RM1	0.7235		0.7196	0.7317	0.7246	0.7245	0.7329	0.7334	0.7278	0.0057	0.7796	0.592
RM2	0.6991	0.7018	0.6969	0.7029	0.693	0.696	0.7031		0.6990	0.0042	0.6045	0.021
FRMH	N/A	0.9069	0.9071	0.9047	0.9047	0.8951	0.9048	0.9159	0.9056	0.0061	0.6725	
FRML	N/A	0.7524	0.758	0.7596	0.758	0.7501	0.7562		0.7557	0.0037	0.4890	
H	N/A	0.6513	0.6525	0.6471	0.6546				0.6514	0.0032	0.4851	
FRHL2	N/A	0.6667	0.6657	0.6719					0.6681	0.0033	0.4982	
FRHH	N/A	0.89999	0.9012	0.8962	0.9087				0.9015	0.0052	0.5810	

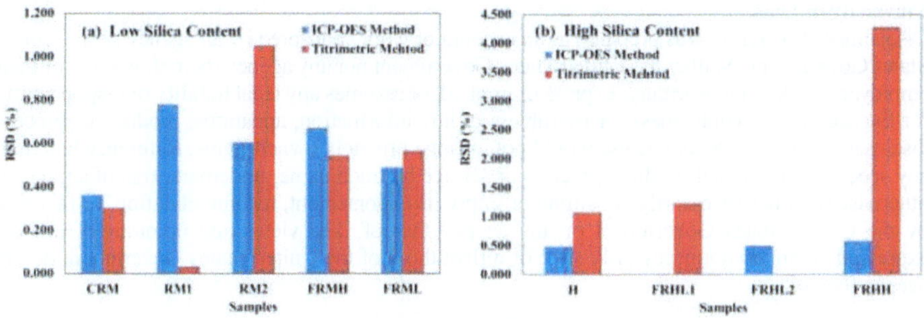

Figure 1. RSD values for the ICP-OES and titrimetric methods for samples with (a) low and (b) high silica contents.

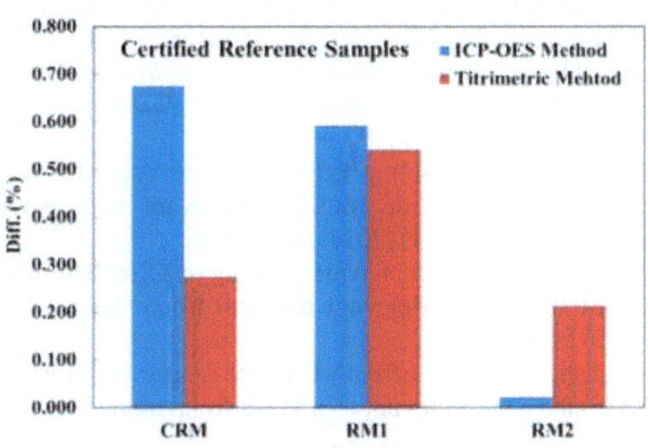

Figure 2. Comparison of the accuracy of both methods.

Conclusions

Titrimetric analysis method was compared to an ICP-OES method developed in this laboratory for determining the total iron fraction in iron ores and DRI samples. The two analysis methods were of similar accuracy and precision. The ICP-OES method is much faster and easier to use than the titrimetric method. The ICP method can be easily used for high or low silica-containing iron samples. Titrimetric method introduces error when analyzing samples with high silica content due to the difficulty in using HF.

Acknowledgments

The authors thank Andrew Laroche for his help with the analytical work using ICP and titration. The authors acknowledge the financial support from the U.S. Department of Energy under Award Number DEEE0005751 with cost share by the American Iron and Steel Institute (AISI) and the University of Utah.

References

1. M.Y. Mohassab-Ahmed and H. Y. Sohn, *Method and Device for Digestion of Materials in a Microwave Oven*, 2012, US Patent App. 61/651.
2. M.Y. Mohassab-Ahmed, H. Y. Sohn, "Effect of Water Vapor Content in H_2–H_2O–CO–CO_2 Mixtures on the Equilibrium Distribution of Manganese between CaO–MgO_{sat}–SiO_2–Al_2O_3–FeO–P_2O_5 Slag and Molten Iron." *Steel Res. Int.*, 85 (2014), 875-884.
3. M.Y. Mohassab Ahmed, "*Phase Equilibria between Iron and Slag in CO/CO_2/H_2/H_2O Atmospheres Relevant to a Novel Flash Ironmaking Technology*" (PhD Dissertation, The University of Utah, 2013).
4. M.Y. Mohassab-Ahmed, H. Y. Sohn, "Effect of Water Vapor Content in H_2-H_2O-CO-CO_2 Mixtures on the Activity of Iron Oxide in Slags Relevant to a Novel Flash Ironmaking Technology." *Ironmaking Steelmaking*, 41(2014), 665 - 675.
5. Y. Mohassab and H. Sohn, "Effect of Water Vapour on Distribution of Phosphorus between Liquid Iron and MgO Saturated Slag Relevant to Flash Ironmaking Technology." *Ironmaking Steelmaking*, 41 (2014), 575-582.
6. Y. Mohassab and H.Y. Sohn, "Effect of Water Vapor on Sulfur Distribution between Liquid Fe and MgO!Saturated Slag Relevant to a Flash Ironmaking Technology." *Steel Res. Int.*, 86 (2014), 753–759.

7. Y. Mohassab and H.Y. Sohn, "Analysis of Slag Chemistry by FTIR!RAS and Raman Spectroscopy: Effect of Water Vapor Content in H_2-H_2O-CO-CO_2 Mixtures Relevant to a Novel Green Ironmaking Technology." *Steel Res. Int.*, 86 (2014), 740-752.
8. M.Y. Mohassab-Ahmed, H.Y. Sohn, and L. Zhu, "Effect of Water Vapour Content in H_2-H_2O-CO-CO_2 Mixtures on MgO Solubility in Slag under Conditions of Novel Flash Ironmaking Technology." *Ironmaking Steelmaking*, 41 (2014), 575-582.
9. F. Chen, Y. Mohassab, T. Jiang, and H.Y. Sohn, "Hydrogen Reduction Kinetics of Hematite Concentrate Particles Relevant to a Novel Flash Ironmaking Process." *Metall. Mater. Trans. B*, 46 (2015), 1133-1145.
10. F. Chen, Y. Mohassab, S. Zhang, and H.Y. Sohn, "Kinetics of the Reduction of Hematite Concentrate Particles by Carbon Monoxide Relevant to a Novel Flash Ironmaking Process." *Metall. Mater. Trans. B*, 46 (2015), 1716-1728.
11. H. Hu, Y. Tang, H. Ying, M. Wang, P. Wan, and X.J. Yang, "The effect of copper on iron reduction and its application to the determination of total iron content in iron and copper ores by potassium dichromate titration. " Talanta, 125 (2014) 425-31.

EPD Congress 2016
Edited by: Antoine Allanore, Laura Bartlett, Cong Wang, Lifeng Zhang, and Jonghyun Lee
TMS (The Minerals, Metals & Materials Society), 2016

EFFECTIVE INOCULATION OF GREY CAST IRON

Dariusz Kopyciński[1], Józef Dorula[2]

[1]AGH University of Science and Technology; 23 Reymont Str.; Krakow, 30-059, Poland
[2]Vesuvius Poland - Foseco Plant, Leonardo da Vinci Str., Gliwice, 44-109, Poland

Keywords: Inoculation, Grey Cast Iron, White Cast Iron, Primary Austenite

Abstract

The study proves that by introducing the iron powder and disintegrated steel scrap to low-sulphur cast iron still before the inoculation carried out with a conventional graphitising inoculant, the mechanical properties similar to those obtained during the inoculation treatment carried out on cast iron with the recommended high sulphur content are achieved. The said operation increases the number of crystallisation nuclei for dendrites of the primary austenite. In this case, the iron particles act as substrates for the nucleation of primary austenite, due to a similar crystallographic behaviour of the regular face centred cubic lattice The more numerous are the dendrites of primary austenite, the less free space is available in the interdendritic spaces for the formation of graphite eutectic grains, which makes the structure more refined (more eutectic grains) and the mechanical properties higher.

Introduction

In industrial practice, during manufacture of iron castings, the inoculation of cast iron consists in introducing the inoculant into liquid metal, which has a low ability to generate the grain nucleation process. The low-weight inoculant introduced during metallurgical treatment improves the molten metal ability to start the nucleation process of graphite grains. Due to the increased number of active substrates for the graphite nucleation, the structure of cast iron is refined and the consequence of this refinement are higher properties of iron castings. From this definition it follows that with the inoculation treatment are closely related some effects that can be evaluated according to selected criteria. Undoubtedly, the most important indicator of the inoculation process effectiveness used in the technology of cast iron manufacture is the increased number of eutectic grains. Another factor taken into account after the inoculation treatment includes a set of characteristic changes that take place in graphite precipitates and are also subjected to evaluation [1,2]. In the structure of inoculated cast iron, the flake graphite of an even distribution is formed. This effect takes place at the cost of graphite characterized by interdendritic distribution which occurs in the base cast iron before inoculation. The value of the supercooling degree ΔT decreases during the crystallization of graphite eutectic in a way similar as the cast iron chill tendency, while the dominant constituent in cast iron microstructure becomes a pearlitic metal matrix. It has been proved that all changes introduced by the inoculation treatment to cast iron microstructure lead to an increase of the mechanical properties. It is worth noting that there is another important indicator of the successful course of inoculation, and at the same most difficult in practical evaluation, and it is the character of changes in the formation of grains of the dendrites of primary austenite [3,4]. This effect is important since the inoculation of grey cast iron affects not only the grains of graphite eutectic, but also the grains of primary austenite. Technical literature states at least four hypotheses on the cast iron inoculation process, but it appears that the hypothesis that has the highest rationale in the industrial practice of cast iron manufacture with ferrosilicon inoculation and minor additions of elements from group II of the periodic table (Ca, Ba, Sr, etc.) and aluminium, is the hypothesis developed by B. Lux [5]. B. Lux in his work [5] has proved that adding the inoculant to cast iron melt leads to the crystallization of carbides with ionic bonds of the CaC_2, BaC_2, BiC_2, SrC_2, and Al_4C_3 type. These carbides act as substrates for the nucleation of graphite.

Moreover, applying B. Lux hypothesis, it becomes possible to evaluate the positive practical aspect of the use of complex ferrosilicon-based inoculants (Fe-Si) composed of silicon (approx. 75%) with small additions of simple compounds introduced in an amount of up to several percent by weight and iron as a remainder. Iron in the inoculant can form a separate phase of the $FeSi_2$ type.

Therefore, it can not be ruled out that this phase can play the role of a proper substrate for the nucleation of the grains of the primary austenite dendrites. It is also worth noting that complex inoculants used in the technology of the manufacture of inoculated cast iron are sometimes enriched with elements such as Bi, Al, La and other rare earth metals, as well as Ti, and without any doubt, during metallurgical treatment, all these elements can form in the liquid metal proper substrates for the nucleation of the grains of the primary austenite dendrites. Detailed analysis of the literature [1,2,5,6] and industrial practice of the cast iron engineering show us that in the process of inoculation, the problem most often considered is the impact of reagents on the grains of graphite eutectic, while possible effect of this treatment on the primary crystallization, that is, on the primary austenite grains, has been so far largely overlooked. Learning the rules of the crystallization of the primary austenite grains in grey cast iron is critical for conscious application of the optimal inoculation treatment. This paper discusses the possibility to control the primary structure formation in cast iron.

Problem of low sulphur content in cast iron

In recent years, the foundry industry has faced numerous transformations that, among others, include changes in the cast iron melting process (departure from cupolas in favour of electric induction furnaces), significant reduction of pig iron content in the charge used for cast iron melting, and increase in the production of ductile iron castings. It should be noted that in the manufacture of ductile iron the basic requirement is low sulphur content. In contrast, in the inoculated cast iron, the required level of sulphur is 0.05% - 0.08%. Therefore foundries which specialize in the ductile iron technology are not able to satisfy the required sulphur content in the inoculated cast iron. This leads to problems with obtaining the regulatory strength parameters in inoculated iron castings and to the occurrence of casting defects of the microporosity type. Figure 1 shows the tensile strength UTS obtained in the inoculated cast iron grade EN-GJL-250 with standard and low sulphur content.

From the above comparison it follows that in the cast iron with low sulphur content it is not possible to comply with the strength requirements specified by standards for a given cast iron grade, in this case $UTS_{min.}$ = 250 MPa. It should be emphasized that in both cases the microstructure of cast iron is similar and characterized by a pearlitic matrix with evenly distributed interdendritic graphite.

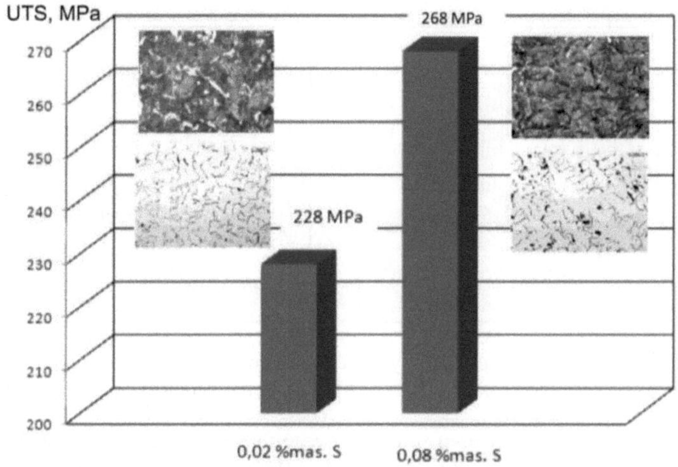

Figure 1. The impact of sulphur in the inoculated grey cast iron on its microstructure and mechanical properties

Moreover, in castings made from the inoculated iron with low sulphur content, the defects of a porosity type may appear. It has been observed that in ductile iron castings, the large number of fine graphite eutectic grains is not favourable in terms of the occurrence of porosity (or shrinkage depressions). On the other hand, increasing the number of primary austenite grains, and thus reducing their dimensions, counteracts this effect. The occurrence of casting defects of this type is best traced during the casting crystallization modelling. Modelling with the PROCAST software is shown on the example of cast electric motor housing. The results of modelling are depicted in Figures 2 and 3.

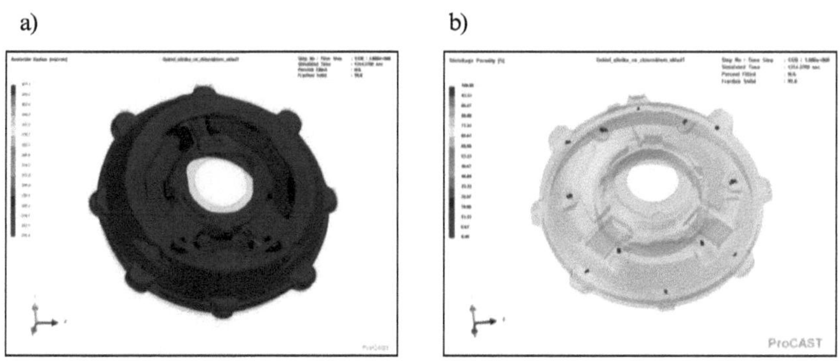

Figure 2. The distribution of primary austenite grains in casting with a small number of substrates for nucleation (a), and microporosity defects visible in this casting (b)

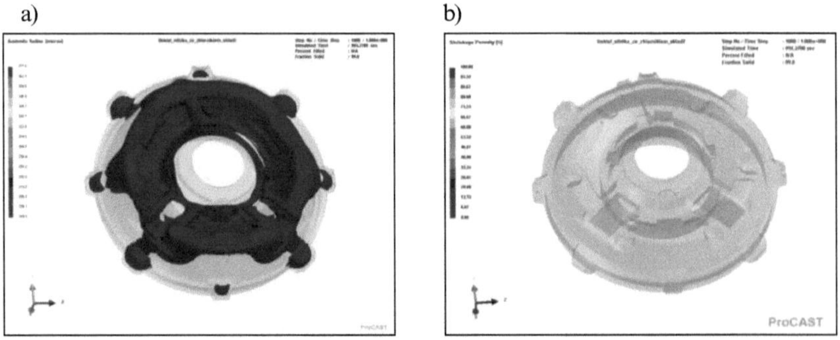

Figure 3. The distribution of primary austenite grains in casting with a large number of substrates for nucleation (a), and absence of microporosity defects in this casting (b)

In simulation, a relationship has been sought between the size of primary austenite grains and possible occurrence of porosity. PROCAST program allows control of a parameter responsible for the average number of substrates for the nucleation of primary austenite grains. This provides a link between the number of substrates for the nucleation of austenite grains and actual number of these grains, allowing also for predicting of their dimensions.
Figure 2a shows the results of computer simulation in which the distribution of primary austenite grains, assumed to have a diameter larger than 250 m, was determined. The simulation was carried out for the case of a small number of substrates for the nucleation of these grains. Figure 2b presents the distribution of shrinkage porosity, which is consistent with the porosity occurring in real casting of the electric motor housing. Figure 3a shows the results of similar modelling of the distribution of primary austenite grains, also assumed to have a diameter larger than 250 m, but this time for the case of a large number of substrates for the nucleation of these grains. The resulting casting shown in Figure 3b is free from the defect of shrinkage porosity. The results of both computer simulations confirm the thesis that during the inoculation, very important is also the size of primary austenite grains, but translating this analysis to the conditions of current foundry production is extremely difficult. In this study it has been proved that under laboratory conditions, the number of primary austenite grains can be increased and their morphology can be changed owing to the effect of common FeSi75 inoculant enhanced with the addition of iron powder or shredded steel scrap (chips, scrap metal, shot, etc.).

Methodology

Test melts were carried out in a medium frequency induction furnace with crucible of 15 kg capacity. The chemical composition of produced cast iron and its mechanical properties are compared in Table 1. Thermal analysis was performed with the ITACA software.
Thus prepared cast iron was subjected to an inoculation method using modifier based on Fe-Si and two step inoculation with the iron powder and inoculant based on Fe-Si. The total amount of inoculant in both cases was 0.4%.

Table 1. Type of inoculant, and chemical composition and mechanical properties of grey cast iron

No.	Metallurgical treatment	Chemical analysis, wt%					S_c^*	UTS [MPa] Mean from three measurements
		C	Si	Mn	P	S		
1.	Base cast iron	2.92	1.65	0.38	0.03	0.014	0.78	Hard spots in sample
2.	Cast iron inoculated with 0.4% Fe powder	2.91	1.66	0.37	0.04	0.013	0.78	290
3.	Cast iron inoculated with 0.2% Fe powder and 0.4 % FeSi75	2.94	1.80	0.39	0.05	0.012	0.79	315

*S_c – degree of eutectic saturation.

Were carried out three melts and made casts of rollers shown in Fig. 4. First melt (no. 1) was without adding inoculant and flooded to form obtain a castings of rollers.
During the second melting (no. 2), the liquid metal has been overheated to a temperature of 1490°C and can hold out for 100 seconds. After lowering the temperature to 1460°C was introduced Fe powder inoculant, waited 180 seconds and in temperature of 1410°C was flooded to form obtain a castings of rollers. In the case of the third melt (no. 3) melting procedure was similar but it was introduced also Fe-Si based before iron powder inoculant. During casting solidification cooling curves were measured. Three melts were carried out and rollers shown in Fig. 4 were cast. Moulds (250mm x 250mm x 170mm) were prepared in No-Bakefuran resin process. When cast in the form of temperature reached 950 °C, a cast was shaken out from mold and placed in a furnace for 30 minutes. After the specified time frame castings were pulled from the furnace and placed in molten salt for isothermal quenching at a temperature of 360°C. (DAAS method [7,8]). Then, it was made test specimens for metallographic.

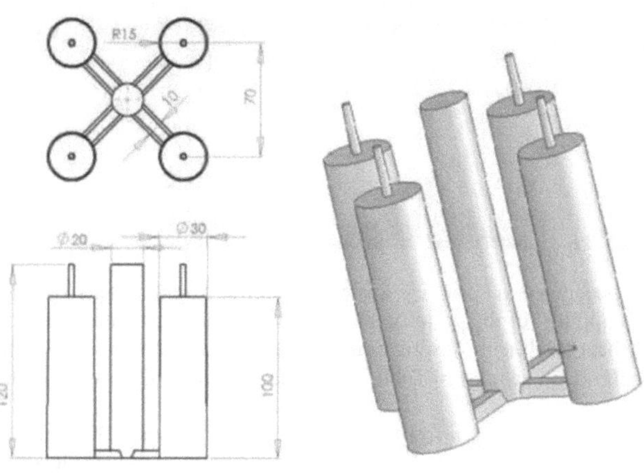

Fig. 4. Scheme of castings of rollers

Evaluation of the inoculation effect

As follows from the tests shown in Figure 5, inoculation treatment changes in the cast iron microstructure not only the number of eutectic grains but also, as has been proved, the number and morphology of primary austenite grains. Therefore, often found differences in the tensile strength UTS have no relevance with the cast iron microstructure revealed by common techniques. In most cases, the cast iron microstructure (without DAAS heat treatment) is as shown in Figure 1.

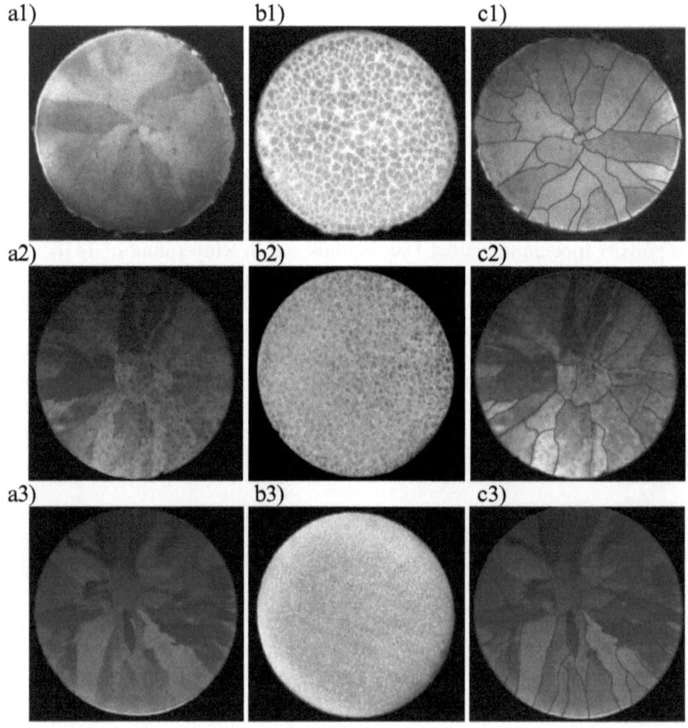

Figure 5. The appearance of primary austenite grains (a) and grains of graphite eutectic (b); visible is also "mapping" of the grain boundaries of primary austenite in sample of grey cast iron after the DAAS heat treatment; designations: 1 – applies to melt no. 1, 2 – applies to melt no. 2, 3 - applies to melt no. 3, as used in Table 1; the actual sample diameter is ϕ 30 mm

Since tensile specimens are taken from the central part of standard 30 mm diameter ingots, only changes in the number of primary austenite grains can explain the tensile strength UTS increase of 25 ÷ 40 MPa. In the middle part of sample No. 3 (Fig. 5-c3), which is the surface of tensile

specimen, the grains of primary austenite are much more numerous and their shape resembles more the shape of equiaxed grains than in the sample shown in Figure 5-c2.
Interesting results of the analysis of primary grains in cast iron are achieved with the ITACA advamced thermal analysis software. ITACA is an abbreviation of the name “Incremental Thermal and Chemical Analysis”. The system was developed in Italy by ProService Technology and distributed in Poland by FOSECO. Figure 6 shows "screenshots” of ITACA system during the thermal analysis conducted for low-sulphur grey cast iron.

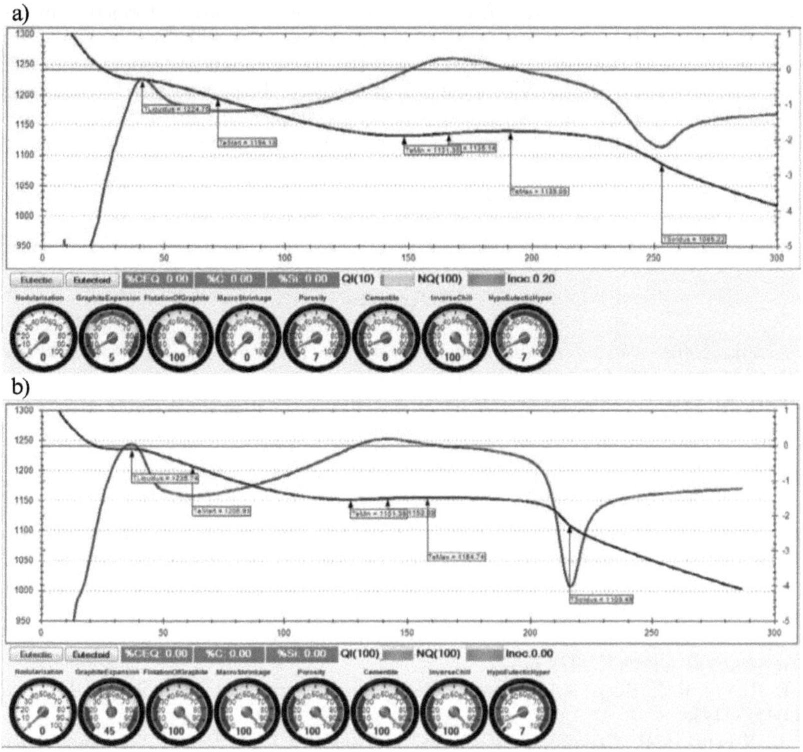

Figure 6. The course of crystallization and cooling of cast iron (Table 1) from melt no. 1 (a) and from melt no. 3 (b), visible is also the graphics of ITACA program displayed in the form of trend indicators, e.g. porosity in casting ("POROSITY" trend indicator, the fourth from the right)

From the course of the crystallization and cooling curves shown in Figure 6, the basic parameters of crystallization were determined by thermal analysis. In this system, worth noting is the graphics of the ITACA program which, among others, allows predicting the porosity in castings ("POROSITY" trend indicator).

Conclusions

Studies prove that it is possible to disclose the primary austenite grains in grey cast iron and link them to the mechanical properties of casting. From the research conducted currectly it follows that the inoculation of grey cast iron characterized by low sulphur content should be assisted with the use of iron particles, which act as substrates for the nucleation of primary austenite grains. Under industrial conditions, the inoculant can be enriched with crushed steel scrap. Then the obtained cast iron will yield products satisfying the normative requirements of mechanical properties. Moreover, as is apparent from Figure 6, the risk of porosity formation in castings drops to zero when the inoculation treatment of grey cast iron with low sulphur content is enhanced by the use of iron particles (Figs. 5c and 6b). These results are certainly valid also for the white cast iron crystallization, where the task of revealing the primary structure is not an easy one at all. However, a comprehensive study of various mechanisms that govern the growth of crystals in white cast iron requires simple methods for the disclosure of primary austenite grains linked to the well visible grains of carbide eutectic, as discussed in [3].

It seems that the condition of iron casting surface (the number of primary austenite grains) will affect the quality of zinc coating applied by the hot-dip galvanization process [9].

References

1. E. Guzik, *Some selected problems concerning the processes of cast iron improvement* (Katowice, Archives of Foundry 1M, 2001), 1-128.
2. E. Fraś, M. Górny, "Inoculation effects of cast iron", *Archives of Foundry Engineering* 12 (2012), 39-46.
3. D. Kopyciński, "The inoculation of white cast iron". *TMS 2013 (The Minerals, Metals & Materials Society). Published by John Wiley&Sons, Inc., Hoboken, New Jersey. Supplemental Proceedings* (2013), 601-608.
4. D. Kopyciński, E. Guzik, A. Szczęsny, "Equiaxed and oriented microstructure in high chromium cast iron", *Archives of Metallurgy and Materials* 59 (2014), 723-727.
5. B. Lux, "Nucleation and graphite in Fe-C-Si alloys". *Recent Research on Cast Iron. Gordon and Breach Publishers*. New York – London – Paris (1968), 241-279.
6. A. Dioszegi, K.Z. Liu, J.L. Svensson, "Inoculation of primary austenite in grey cast iron", *Cast Metals Research* (2007), 68-72.
7. G. L. Rivera, R. E. Boeri, J.A. Sikora, "Solidification of grey cast iron", *Scripta Materialia* 50 (2004), 331-38.
8. G. L. Rivera, P. R. Calvillo, R. E. Boeri, Y. Houbaert., J.A. Sikora, "Examination of the solidification macrostructure of spheroidal and flake graphite cast irons using DAAS and ESBD", *Materials Characterization* 59 (2008), 1342-46.
9. D. Kopyciński, "Crystallization of intermetallic phases Fe-Zn during hot-dip galvanizing process", *TMS 2013 (The Minerals, Metals & Materials Society). Published by John Wiley&Sons, Inc., Hoboken, New Jersey. Supplemental Proceedings* (2013), 439-446.

EPD Congress 2016
Edited by: Antoine Allanore, Laura Bartlett, Cong Wang, Lifeng Zhang, and Jonghyun Lee
TMS (The Minerals, Metals & Materials Society), 2016

Study on the Infrared Spectral Range for Radiation Temperature Measurement of Continuous Casting Slab

Yunwei Huang, Dengfu Chen*, Lin Bai, Mujun Long, Kui Lv, Pei Xu

Laboratory of Metallurgy and Materials, College of Materials Science and Engineering, Chongqing University, Chongqing 400030, P.R. China.
Email: chendfu@cqu.edu.cn

Keywords: Radiation temperature measurement, Continuous casting, Infrared spectral wavelength

Abstract

The surface temperature of continuous casting slab in the secondary cooling zone is a key parameter of production control. However, the accuracy of radiation temperature measurement of high temperature casting slab is greatly affected by the environmental factors, such as water vapor, spray, water film and other participational medias during continuous casting process. In this paper, the radiation characteristics of water vapor and its influences to radiation temperature measurement of continuous casting slab were investigated. Based on the database HITRAN 2004 and the measuring principle of monochrome thermometer & colorimetric thermometer, the measuring errors under different operating infrared spectral wavelength were compared. Results showed that spectral region with wavelength nearby 1.0 m or 1.6 m would be the optimum for high temperature infrared thermometer used under the conditions with moist steam, water spray and water film at the secondary cooling zone in steel continuous casting.

Introduction

The temperature measurement of continuous casting slab is of importance for process control and product quality control, but its value can't be measured accurately. Water vapor is the main component of the infrared radiation in the secondary cooling zone. Its spectroscopy is very complex[1] and the radiation characteristic spectral data has been studied theoretically and experimentally for many years abroad[2]. But regarding the temperature range for present industrial usage at secondary cooling zone, the data was inadequate, low-resolution and inaccuracy in some literature[3]. So for the radiative transmission characteristics and the revision of surface infrared temperature measurement of the continuous casting slab, it's valuable to expand the water vapor spectral band model parameters. HITRAN[4] is a database of high-resolution transmission molecular absorption to predict and simulate the transmission and emission of light. For water vapor, there are 64 thousands spectral lines parameters in HITRAN. Such a big database is extraordinarily inconvenient and the computation is time-consuming. Therefore we need to create a simple database which can describe the water vapor spectral characteristics with particular precise.

According to the conditions for steel continuous casting, the infrared radiation spectral band-model parameters for water vapor were studied and expanded. Based on the database HITRAN 2004, the article calculated the narrow-band model parameters (average absorption factor $\bar{\kappa}_\eta$, spectral line density $1/\bar{d}$, spectral line half-width $\bar{b}_L$) in the great spectra region and temperature range for the calculation application of infrared radiation transfer. With the measuring principle of monochrome thermometer, colorimetric thermometer, the measuring errors under each operating infrared spectral wavelength were calculated.

Theoretical formulation

Considering a great number of water vapor bands lines, the position of spectral line in the band was supposed to be randomly distributed and the intensity of spectral line conforms to the statistic intensity distribution rule. If the spectral line intensity distribution conforms to the indices-caudal reciprocal distribution, the average penetration can be presented by:

$$\bar{\gamma}_\eta = \exp\left[-\frac{2\bar{b}_L}{\bar{d}}\left(\sqrt{1+\bar{\kappa}_\eta \bar{d} X/\bar{b}_L}-1\right)\right] \tag{1}$$

where $X = P_{H_2O}L(298K)$ is pressure path, which should be standardized to 0.1 MPa, 298K; P_{H_2O} is water vapor partial pressure in Pa; L is path length in meter.

Based on the database HITRAN, the spectral band model parameters were calculated in the study [5~7]:

$$\bar{\kappa}_{\eta_i} = \frac{1}{\Delta\eta_i}\sum_{m=1}^{M} S_i^m \tag{2}$$

$$\bar{b}_{L_i} = \frac{1}{M}\sum_{m=1}^{M} b_{L,i}^m \tag{3}$$

$$\bar{d}_i = \frac{\bar{\kappa}_{\eta_i}\bar{\gamma}_{L_i}}{\left(\frac{1}{\Delta\eta_i}\sum_{m=1}^{M}\sqrt{S_i^m b_{L,i}^m}\right)^2} \tag{4}$$

Where S_i^m is the intensity of spectral line m at region i in (cm·MPa)$^{-1}$; M is the spectral line total at region i; $\Delta\eta_i$ is the wave number interval at region i in cm^{-1}.

Monochromatic infrared temperature measurement

Most of the infrared thermometers are based on the brightness. The target signal measured by the infrared thermometer is the brightness temperature. At a certain wavelength, if the emissive energy from the actual body with blackness of ε_λ is the same as that from an ideal black body, the temperature of the black body is supposed to be the brightness temperature (T_S) of the actual body. The relationship between the brightness temperature and the actual temperature (T) of the actual body can be expressed by:

$$\frac{1}{T_S} - \frac{1}{T} = \frac{\lambda}{C_2} \ln \frac{1}{\varepsilon_\lambda} \tag{5}$$

Then the actual temperature T of the actual body can be calculated by:

$$T = \frac{T_S C_2}{\lambda T_S \ln \varepsilon_\lambda + C_2} \tag{6}$$

ε_λ is no more than 1.0, and varies with the wavelength. Ignoring the scattering of the water vapor, the transmissivity of the water vapor can be obtained according to the Kirchhoff's law:

$$\gamma_\lambda = 1 - \varepsilon_\lambda \tag{7}$$

$$\varepsilon_\lambda^{'} = \gamma_\lambda \varepsilon_\lambda \tag{8}$$

Thus the actual temperature measured by the way of monochromatic infrared temperature measurement can be rewritten as:

$$T = \frac{T_S C_2}{\lambda T_S \ln \gamma_\lambda \varepsilon_\lambda^{'} + C_2} \tag{9}$$

Colorimetric infrared temperature measurement

A colorimetric infrared thermometer is supposed to consist of two monochromatic brightness thermometers in the same package. With two detector layers, the colorimetric infrared thermometer simultaneously measures the temperature of the target body at two separate wavebands. Then the two signals are processed as a ratio. As long as the partial obstruction or attenuation affects each of the wavelengths by an equal amount, the ratio of the two signals is accurate enough for the calibration curve. For a non-black target body with temperature T and a black body with temperature T_C, if their ratio of the spectral radiation brightness at λ_1 and λ_2 is the same, T_C is supposed to be the color temperature of the target body. The relationship of the actual temperature T and the T_C of the target body is expressed by:

$$\frac{1}{T}-\frac{1}{T_C}=\frac{\ln\frac{\varepsilon_{\lambda_1}}{\varepsilon_{\lambda_2}}}{(\frac{1}{\lambda_1}-\frac{1}{\lambda_2})\cdot C_2} \tag{10}$$

Then the actual temperature of the target body can be calculated by:

$$T=\frac{T_C}{1+T_C\ln\frac{\varepsilon_{\lambda_1}}{\varepsilon_{\lambda_2}}\Big/\left[(\frac{1}{\lambda_1}-\frac{1}{\lambda_2})C_2\right]} \tag{11}$$

Where $\varepsilon_{\lambda 1}/\varepsilon_{\lambda 2}$ is the emissivity ratio at the two wavebands, which is called the gradient of the colorimetric thermometer.

Results and discussions

According to the environment conditions of secondary cooling zone in steel continuous casting, the effect of water vapor on the infrared radiation spectral band-model parameters was studied. Table 1 lists the parameters used in the calculation. With Eqs. (2)-(4), the parameters of narrow-band model $\overline{\kappa_{\eta_i}}$, $\overline{\gamma_{L_i}}$ and. $\overline{d_i}$ were calculated, and a small parameter database was set up for the infrared radiation spectral at secondary cooling zone in steel continuous casting. The data can be used to amend the effect of vapor absorption on the infrared thermometer.

Table 1 Parameters used in the calculation

Pressure	Temperature range	Interval of Temperature	wave number region	Interval of wave number
0.1 MPa	300-2000 K	50 K	700-10000 cm^{-1}	5 cm^{-1}

Spectral absorbance of saturated water vapor at different conditions

With the parameters in table 1 and other parameters of narrow-band model serving as a basis, spectral emissivities (absorbance) of Saturated water vapor at 500K, 0.1MPa L=100cm were calculated in different wavenumbers. The spectral regions are 8-14 m, 5.0 m, 1.6 m, 1.0 m as shown in Fig.1 (a)-(d). The results indicated that spectral range about 5.0 m and 8-14 m is not suitable for thermometer because of its large absorbance and fluky changes. And the spectral absorbance of at the range nearby 1.6 m and 1.0 m is less than 0.005. They are almost "translucent". Table 2 lists a part of the results under wavelength of 1.587-1.613 m, which is commonly used by portable Infrared thermometer in the metallurgical industry.

Table 2. Parameters of narrow-band model in the 6200 to 6280 cm^{-1}

Wavenumber	300 K			400 K			500 K		
	κ	1/d	γ	κ	1/d	γ	κ	1/d	γ
6202.5	1.19E-07	0.041962	0.67546	3.61E-07	0.036134	0.40402	7.50E-07	0.032752	0.20658
6207.5	9.73E-07	0.043668	0.79222	8.35E-07	0.037518	0.38028	1.14E-06	0.033922	0.19637
6212.5	8.86E-08	0.043641	0.46336	3.21E-07	0.037465	0.30610	1.10E-06	0.033857	0.23206
6217.5	1.32E-06	0.043758	0.38066	1.35E-06	0.037587	0.23976	2.05E-06	0.033979	0.16496
6222.5	2.73E-07	0.044190	0.98971	4.68E-07	0.037900	0.52477	8.77E-07	0.034217	0.26692
6227.5	2.56E-07	0.044181	0.51074	5.83E-07	0.037884	0.34077	1.11E-06	0.034198	0.19443
6232.5	1.68E-07	0.045403	0.52615	2.50E-07	0.038918	0.23150	6.51E-07	0.035100	0.15115
6237.5	1.07E-07	0.043715	0.38730	3.57E-07	0.037542	0.22552	1.17E-06	0.033937	0.17304
6242.5	1.92E-06	0.044974	0.64298	1.42E-06	0.038529	0.38072	1.59E-06	0.034746	0.20047
6247.5	2.74E-07	0.045749	1.24510	5.11E-07	0.039166	0.58864	8.10E-07	0.035289	0.24292
6252.5	5.65E-07	0.047493	0.88831	6.19E-07	0.040598	0.41075	8.21E-07	0.036511	0.16626
6257.5	9.27E-07	0.045599	0.44390	8.53E-07	0.039057	0.24519	1.13E-06	0.035206	0.13559
6262.5	3.16E-06	0.043782	0.38651	2.42E-06	0.037616	0.24488	2.39E-06	0.034015	0.13026
6267.5	4.36E-07	0.046252	0.49402	7.19E-07	0.039562	0.29688	1.28E-06	0.035619	0.19068
6272.5	2.12E-06	0.046071	0.63287	2.35E-06	0.039416	0.37975	3.16E-06	0.035495	0.21918
6277.5	1.34E-06	0.047208	0.65398	1.01E-06	0.040336	0.32693	1.20E-06	0.036271	0.15697
6282.5	1.63E-06	0.042984	0.60455	1.96E-06	0.036973	0.28471	2.99E-06	0.033480	0.15919

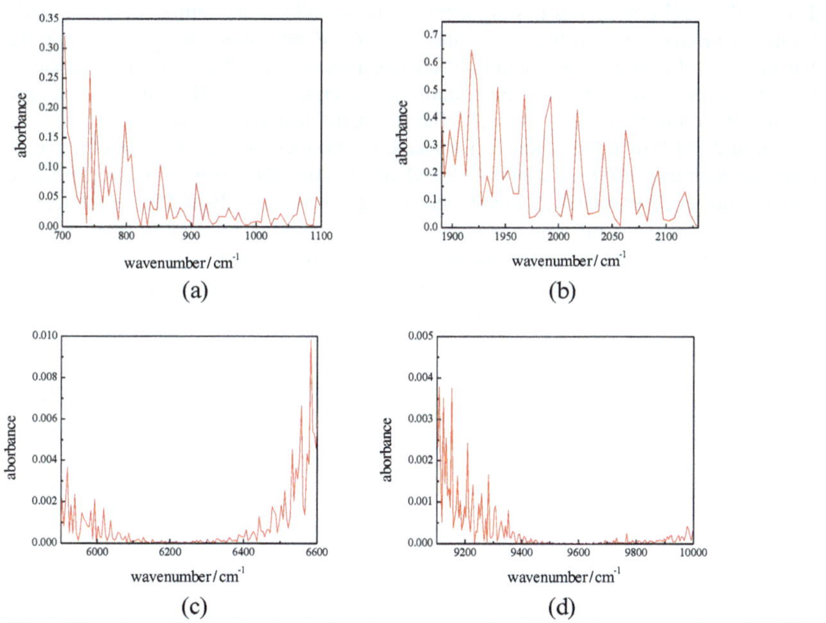

Fig.1. The absorptions of saturated water vapor under different wavenumbers (a) 700–1100cm^{-1}, (b) 1900-2100cm^{-1}, (c) 6100-6600cm^{-1}, (d) 9100-10000 cm^{-1}.(P=0.1Mpa, T=500K, L=100cm)

Fig.2 and Fig.3 are the absorptions comparison of saturated water vapor(P=0.1Mpa, wavelength is 1.6 m) under different conditions. They indicated it has no remarkable effect on the infrared radiative spectral characters when temperature or measuring distance changes. when the wavelength nearby 1.6 m, the absorptions of saturated water vapor is small and steady, so the thermometer is a good choice to be used at secondary cooling zone of continuous casting.

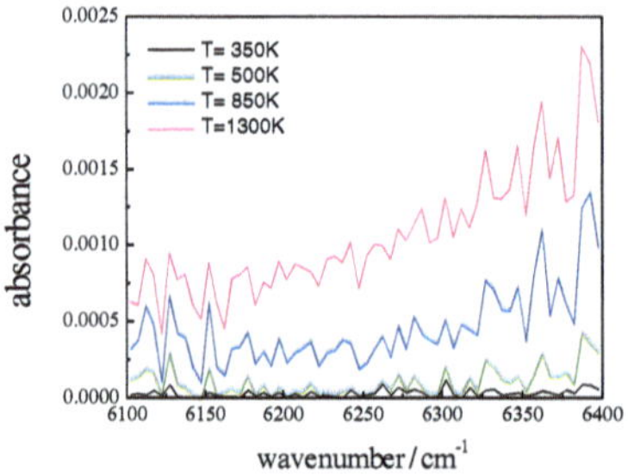

Fig.2. The absorptions comparison of saturated water vapor at different measuring temperature (P=0.1Mpa, L=100cm, wavelength is 1.6 m)

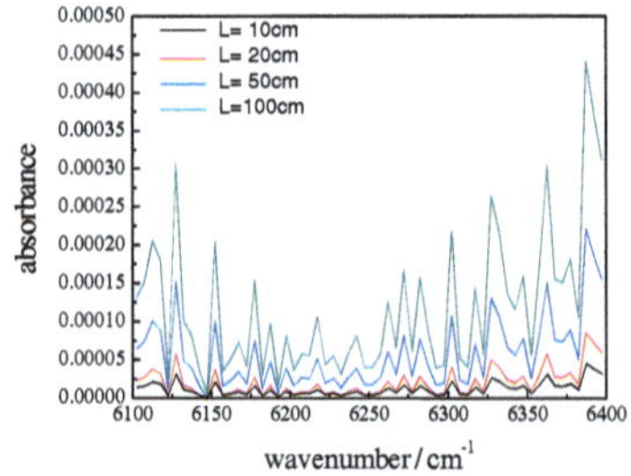

Fig.3. The absorptions comparison of saturated water vapor at different measuring temperature (P=0.1Mpa, T=500K, wavelength is 1.6 m)

Effect of water vapor on the monochromatic infrared temperature measurement

During the steel continuous casting process, the strand surface temperature is usually high. Assuming 0.85 of blackness and 1000°C of brightness temperature for the thermal strand during steel continuous casting process, the effect of water vapor on the monochromatic infrared temperature measurement under different wavelength was illustrated in Figures 4(a) to 4(f). The results indicated that spectral range nearby 5.0 m and 8-14 m is not suitable for usage of monochromatic thermometer for its large temperature difference and fluky changes. And the error of monochromatic temperature measurement at range nearby 1.6 m is less than 2℃. Because the absorbances of spectral range nearby 1.0 m is similar to 1.6 m, the error of monochromatic temperature measurement at range nearby 1.0 m is also very small. Therefor, the spectral range nearby 1.6 m and 1.0 m is suitable for usage of thermometer used at secondary cooling zone of continuous casting.

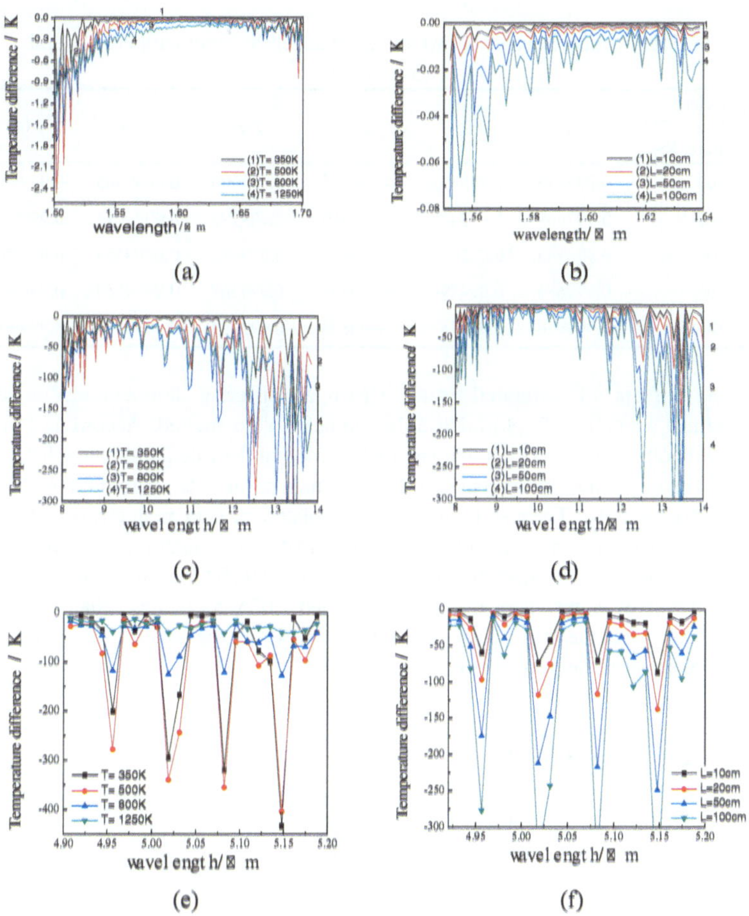

Fig.4. The error for monochromatic temperature measurement under different wavelengths and temperature (a) L=100cm, wavelength is about 1.6 m, (b) T=500K, wavelength is about 1.6 m, (c) L=100cm, wavelength is 8-14 m, (d) T=500K, wavelength is 8-14 m, (e) L=100cm, wavelength is about 5 m, (f) T=500K, wavelength is about 5 m.

<u>Effect of water vapor on the colorimetric infrared temperature measurement</u>

There are rare wavebands used for the colorimetric thermometer. For temperature range of 700-3500 °C, the waveband is 0.7-1.08 m. The waveband is 1.51-1.60 m or 1.65-1.71 m for the temperature range of 250-1000 °C. Most of the thermometers used in metallurgical industry are with 0.95 m wavelength or 1.05 m wavelength. According to the data base above, the transmissivities of water vapor under different temperature and wavelength were obtained (Table 3).

Table 3 The transmissivities of water vapor under different temperature and wavelength

Temperature, K \ Wavelength, m	0.925	0.95	1.025	1.05	0.89	1.08
300	0.999775	0.996373	0.999998	1.000000	0.9999905	0.999907
400	0.997669	0.963428	0.999969	0.999996	0.9999013	0.998215
500	0.998945	0.982616	0.999966	0.999988	0.9999554	0.998783
700	0.999685	0.994294	0.999897	0.999967	0.9999867	0.999368
1000	0.999916	0.998151	0.999873	0.999927	0.9999965	0.999690

Then the effects of saturated water vapor at different temperature on the measurement accuracy of colorimetric thermometer were studied. According to the steel continuous casting process, the parameters used in the calculation were listed in Table 4. The calculated measurement errors versus measured temperature are shown in Figure 5(a) and 5(b) The results shown the calculated measurement errors of MR1S Colorimetric Thermometer (wavelength is 0.95/1.05 m) is larger than FAR Fiber Type Colorimetric Thermometer (wavelength is 0.925/1.025 m), because the transmissivities of water vapor under 0.925 m and 1.025 m is closer than 0.95 and 1.05 m. The closer the transmissivities of two waves are, the smaller the measuring error of colorimetric thermometer is.

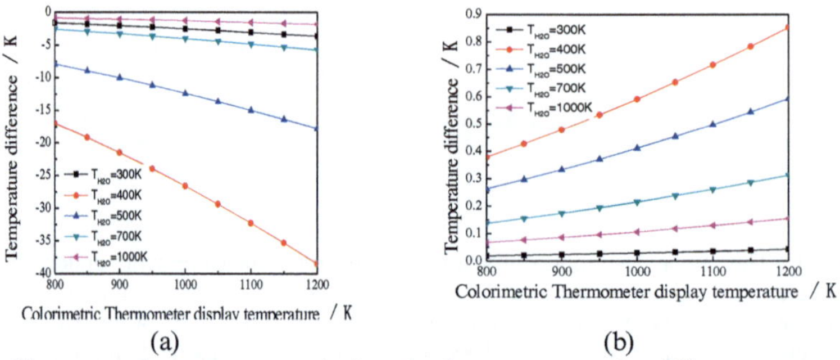

(a) (b)

Fig.5. The measuring errors on colorimetric thermometer under different saturated water vapor temperatures (L = 100cm, gradient suppose to be 1.0) (a) MR1S Colorimetric Thermometer (wavelength is 0.95/1.05 m), (b) FAR Colorimetric Thermometer (wavelength is 0.925/1.025 m).

Table 4 Parameters used in the calculation of colorimetric measurement

Blackness	Measured distance	Wavelength	Colorimetric thermometer gradient
0.85	1.0m	1.05 m	1.0

Summary

Bąsed on the gas spectral data base HITRAN 2004 and the measuring principle of monochrome thermometer & colorimetric thermometer, the water vapor spectral band model was developed and the effect of water vapor on the infrared temperature measurement was studied. The results showed that spectral region with wavelength near 1.0 m or 1.6 m would be the optimum for high temperature infrared thermometer used under the conditions with moist steam, water spray and water film at the secondary cooling zone in steel continuous casting. The closer the transmissivities of two waves are, the smaller the measuring error of colorimetric thermometer is.

Acknowledgements

The work is financially supported by the Natural Science Foundation of China, Project No. 50674109 and 51504048.

References

[1]P. F. Bernath, The spectroscopy of water vapour: Experiment, theory and applications, Phys. Chem. Chem. Phys. 4 (2002) 1501-1509.
[2]Jermy H. Schryber, Steven Miller, and Jonathan Tennyson. Computed Infrared Absorption Properties of Hot Water Vapor. JQSRT, 1995, 53 (4): 373-380.
[3]K. M. Smith, I. Ptashnik, D. A. Newnhama, etal. Journal of Quantitative Spectroscopy & Radiative Transfer(JQSRT) 83 (2004) 735–749.
[4] L. S. Rothman, D. Jacquemart, A. Barbe, et al. The Hitran 2004 molecular spectroscopic database. JQSRT, 2005, 96 (2): 139-204.
[5]YOUNG S J. Nonisothermalband model theory[J]. JQSRT, 1976, 18: [1-28].
[6] FERRISO C C, LUDWING C B. Spectral emissivityes and integrated intensities of the 2.7 mu meter H2O band between 530 and 2200K[J]. JQSRT, 1964, 4: 215-227.
[7] D. Eisenberg and W. Kauzmann, The structure and properties of water[M], Oxford University Press, London, 1969.

EPD Congress 2016
Edited by: Antoine Allanore, Laura Bartlett, Cong Wang, Lifeng Zhang, and Jonghyun Lee
TMS (The Minerals, Metals & Materials Society), 2016

Behavior of Quartz and Carbon Black Pellets at Elevated Temperatures

Fei Li, Merete Tangstad

NTNU (Norwegian University of Science and Technology)
Alfred Getz vei 2, Bergbygget, Trondheim, N-7491,Norway

Keywords: quartz and carbon black pellets, elevated temperatures, reaction kinetics mechanism

Abstract

This paper studies the quartz and carbon black pellets at elevated temperature with varying temperature and gas atmosphere. High-purity quartz and commercial ultra-pure carbon black was mixed (carbon content *wt*.15%), and then pelletized into particles of 1-3mm in diameter. The stoichiometric analysis of the pellet during heating is studied in thermogravimetric analysis (TGA) furnace at different temperature in CO and Ar atmosphere. The microstructure, phase changes and element content of sample before/after heating is characterized by X-ray diffraction, scanning electron microscope, X-ray fluorescence and LECO analyzer. The reaction process can be divided into two stages. Higher temperature and argon atmosphere are the positive parameters for SiC formation.

Introduction

In the silicon production process quartz and carbonaceous materials are the main raw. Silicon is produced in the submerged arc furnace and the temperature inside the furnace ranges from 1300°C to over 2000°C [1,2]. Quartz and carbon can also be used as the raw material for carbothermal reduction synthesis the ceramic material SiC.

Many scientists have studied carbothermal reduction of silica for many years, which is of importance in both the two processes mentioned above. All of publications agree that the reaction will go through the production of gas species, which can be SiO, CO and CO_2. The reaction through gas species is expected from solid-solid reactions. Biernacki[3] has put forth a reaction path mechanism in which the three interrelated reactions occur during the SiO_2/C reaction between 1200 and 1650°C. This study was designed to investigate the 1:1 molar reaction of SiO_2 and C.

$$SiO_2(s.l) + C(s) = SiO(g) + CO(g) \quad (1)$$

$$SiO(g) + 2C(s) = SiC(g) + CO(g) \quad (2)$$

$$2SiO_2(s.\ l) + SiC(s) = 3SiO(g) + CO(g) \quad (3)$$

Agrarwal and Pal[4] studied the behavior of silica and carbon black pellets in argon gas at 1600°C. They found that the following two gas-solid reactions and reaction (2) are the main reactions.

$$SiO_2(s) + CO(g) = SiO(g) + CO_2(g) \quad (4)$$

$$C + CO_2 = 2CO \quad (5)$$

Li et al[5] studied the reduction mechanism of quartz and graphite pellets with the C/SiO_2 mole ratio of 3.6 in the temperature range from 300°C to 1600°C. They concluded that synthesis of SiC from quartz and graphite in pure argon was initiated by reactions (1) and (5) and then followed by the reactions (4), (2) and (5).

It is also generally accepted reaction (6) is the SiC production reaction by the studies of beta SiC powder production via the reduction of SiO_2. Reaction (6) comes out when reaction (1) and (2) are added.

$$SiO_2(s,l) + 3C(s) = SiC(s) + 2CO(g) \tag{6}$$

Selvam et al [6], Koc et al [7], Lin et al [8] and Yao et al [9] found that SiC whiskers or fibers is also produced when they synthesize SiC. The SiC whiskers were generated when SiO vapour reacts with CO vapour by the following reaction:

$$SiO(g) + 3CO(g) = SiC(s) + 2CO_2(g) \tag{7}$$

In order to better understand the reaction mechanism of quartz and carbon pellets at elevated temperature, this paper will focus on the behavior of pellets during thermogravimetric analysis (TGA) furnace at varying holding temperature and atmosphere. Pellets before and after heating will be analyzed by SEM, EDAX, XRD, XRF and LECO.

Experiments and methods

The SiO_2 and the main impurity content of quartz analyzed by XRF are listed in Table 1. The quartz is quite pure with a total of 99.4% SiO_2. Quartz lumps was crushed into powder by a tungsten carbide crusher. Carbon black (Thermax®N991 Powder Ultra-Pure) is used as the carbon source in the pellet. The particle size of quartz and carbon black, shown in Figure 1, was conducted by Laser Diffraction Particle Size Analyzers (COULTER LS 230). The diameter of the quartz was distributed evenly from 1 to 39 m, while distribution of carbon black is from 0.6 to 3 m. Quartz and carbon black powder (carbon black proportion: 15 *wt.* %) were mixed in a mixer for 8 hours. Then the mixture is pelletized into 1-3 mm pellets using water as binder. The mixed quartz and carbon black pellets were cured in a drying oven at 100℃ for 12 hours to vaporize the water in pellets. Figure 2 shows the SEM pictures of the carbon black and the pellets used in the experiments. The morphology of carbon black is the typical spherical particles. The exterior surface and the mid-section of the pellet look quite similar. The carbon black (spherical particles) adheres to the surface of the quartz grain. The quartz grains are typical a non-uniform polyhedron with sharp edges.

Table 1 SiO_2 and main impurities content of quartz (*wt.* %)

Substance	SiO_2	Al_2O_3	Fe_2O_3	K_2O	Na_2O	CaO	TiO_2
Content	99.4	0.096	0.019	0.040	0.004	0.004	0.0009

All the TGA experiments were carried out using thermo-gravimetric furnace (Figure 3) with 0.5 L/min CO or argon gas flow. An electrical balance was placed on the top of the furnace, which was connected to the graphite sample holder by a molybdenum hook. The controlling software records the sample weight every 5 seconds. The tip of the thermocouple is 0.5cm below the bottom of the carbon crucible. Gas flows into the chamber from the top of the furnace and flows out in the bottom. During the experiment, the pressure inside the furnace is approximately 1 bar. About 5g of pellets were used in each experiment. The sample was heated from room temperature with 25℃/min to holding temperature, and then kept for 60 minutes. The cooling time of each experiment from the holding temperature to 1000℃ took about 6 minutes. The crucible used in all experiments has three holes with the diameter of 3 mm in the lid and three holes with the same diameter in the wall of crucible, to ease the gas flow.

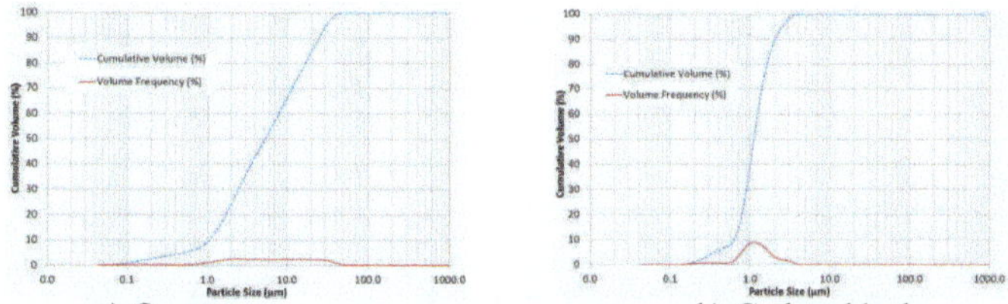

a) Quartz b) Carbon black

Figure 1 Particle size distribution of quartz and carbon black

a) b) c)

Figure 2 SEM picture of carbon black and pellets used for experiments
a) Carbon black b) exterior surface of pellet c) mid-section of pellets

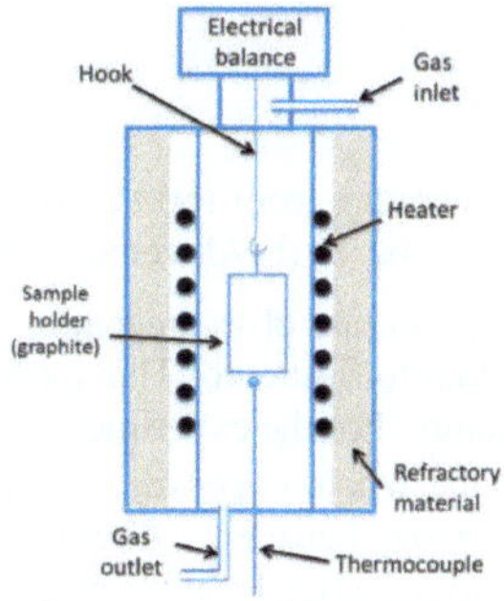

Figure 3 Schematic diagram of thermo-gravimetry (TGA) furnace used in experiment

Phases and structures of the pellets were determined by X-ray diffraction (XRD, Bruker AXS D8 Advance diffractometric with Cu Kα1 radiation). The morphology samples were investigated by field-emission scanning electron microscopy (LVFESEM, Zeiss Supra 55VP) with energy-dispersive X-ray spectrometers (EDX). Si, total carbon, O and free carbon content analysis were conducted by XRF Quantative Analysis, Leco CS600, O Leco TCH600, free carbon Leco Rc-412 respectively.

Results and discussion

A. The effects on the holding temperature

The heating experiments of mixed quartz and carbon black pellets in temperature range from 1600℃ to 1675℃ were conducted in CO gas. The weight loss (%) and weight loss ratio (d%/dt) versus time at different holding temperature is shown in Figure 4.

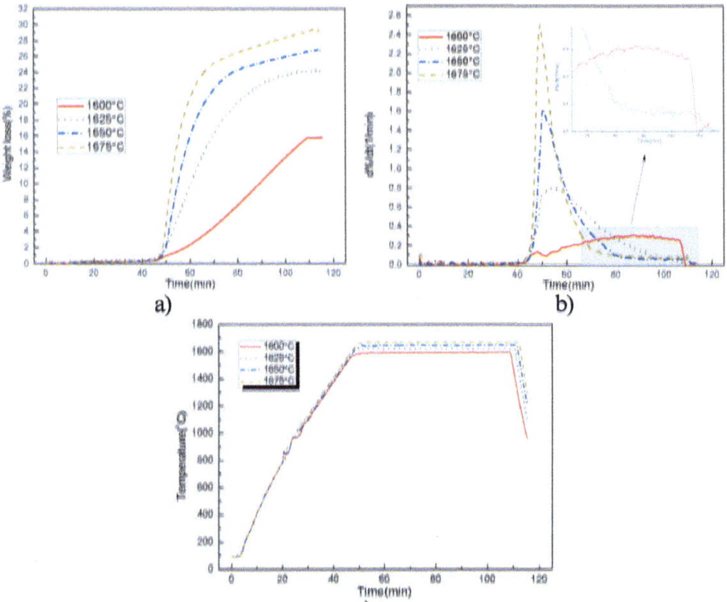

Figure 4 Effect of holding temperature from 1600℃ to 1675℃ in CO gas atmosphere.
a) Weight loss (%) versus time b) d%/dt versus time c) temperature profile

Weight loss ratio (d%/dt) is the reaction rate of the reaction. The peak value of d%/dt increased when temperature increased according to figure 4b. This means the reaction between quartz and carbon black is sensitive to temperature. For the experiment at 1600℃, there is no apparent peak. The reaction rate was low during the whole experiment. The peak at 1625℃ is also not fully developed comparing with the last two experiments. The weight loss rate curves of experiments at 1650℃ and 1675℃ both started to rise after 1550℃. After the peak, the reaction speed started to slow down, then the reaction speed kept at low pace until the end of the experiments. The whole reaction process can be divided into two stages according to the reaction rate, the slow stage and fast stage. The first step of stage 1, which lasts several minutes, from starting point to peak rate (the reaction rate increased); the second step of stage 1, after peak to constant rate (the reaction rate decreased); The second stage, after stage 1 to the end (the reaction rate is stable). In these two stages, different reactions were involved.

The final weight loss of each experiment increased as the holding temperature increased from 1600℃ to 1675℃. As shown in figure 4a, all the weight loss curves of the four experiments started to rise at the time around 46 minutes. The temperature of raising point is around 1550℃ according to figure 4c. This means the reaction between quartz and carbon black in this investigation starts around 1550°C. The curves of weight loss for 1650℃ and 1675℃ also have its' own features. The curves have constant reaction rates at stage 2. The conjunction part from stage 1 and stage 2 for these two experiments are all 2-3% higher than 23%. The value of 23% is the theoretical weight loss when quartz and carbon black reacts according to reaction 7 at carbon content of 15%. It is hence reasonable to believe that the main reaction during stage 1 is reaction 6, that is the main reaction at stage 1 is the production of SiC. Though this reaction has to go through a gas phase, it seems like there is no loss of SiO gas during the first stage, as the turning point of the graph is close to 23%, the theoretical weight loss of CO.

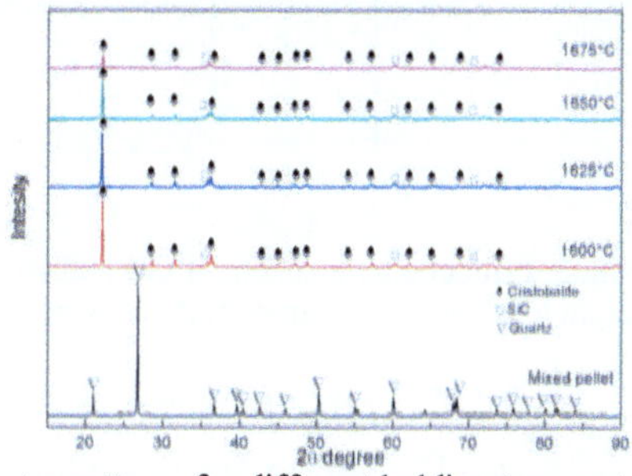

Figure 5 X-ray diffractometer patterns for different holding temperature from 1600°C to 1675°C in CO gas atmosphere and mixed quartz and carbon black pellets (before heating)

Figure 5 presents the phase changes of XRD patterns for four different holding temperatures. Carbon black is difficult to detect in pellets before heating because it is amorphous. When comparing with the mixed quartz and carbon black pellets, quartz and free carbon can't be found in heated samples. The intensity of main peaks for cristobalite is fading while the intensity of SiC increasing when holding temperature increased from 1600°C to 1675°C. This means the quartz phase transferred to cristobalite during experiment and also indicates a higher consumption of SiO_2 at higher temperature. The content of SiC in samples increased as the holding temperature increased. For all the four samples after heating, carbon content is too low to detect. However, the pellets heated at 1600°C, with less transformed carbon to SiC according to the weight loss, has a darker color compared to the other pellets.

B. The effects of gas atmosphere

In order to investigate the effect of gas atmosphere, experiments were conducted in argon and CO atmosphere at 1600°C and 1650°C respectively. The final weight loss of experiments in argon is obviously higher than experiments in CO. As shown in Figure 6a, the starting point of reaction in argon is earlier than CO. This indicates that CO gas is the resultant of stage 1. According to figure 6b, the second stage for experiments at 1650°C in argon and CO last approximately 12 min and 29 min respectively. This duration difference may be caused by the CO supplement to the reactions in this stage. In CO atmosphere, the CO transfers to the reaction interface continuously, while only limited amount of CO generated from stage 1 available in experiment in argon. At stage 2, the reaction rate (figure 6b) in argon atmosphere is two times higher than CO at 1650°C. This may be caused by two reasons: 1) a higher SiC content is produced in argon atmosphere than CO. The pellet's weight loss in argon atmosphere are higher than CO, which means more SiC generated. This makes the contact area of reactant SiO_2 and SiC in stage 2 larger. 2) The CO gas atmosphere depresses reaction 3.

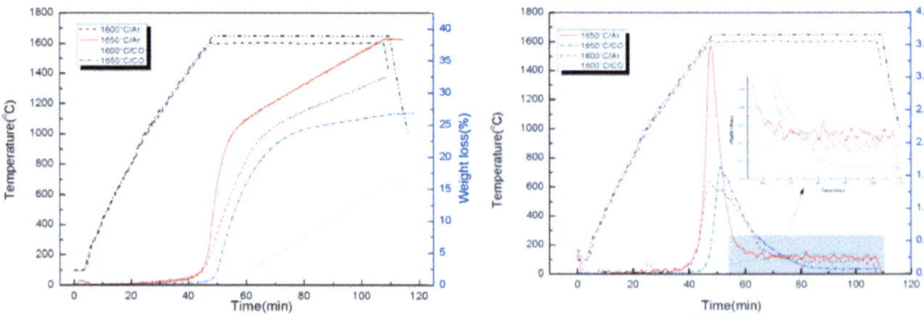

a) b)

Figure 6. Effect on atmosphere of CO and argon.
a) Weight loss (%) versus time b) d%/dt versus time

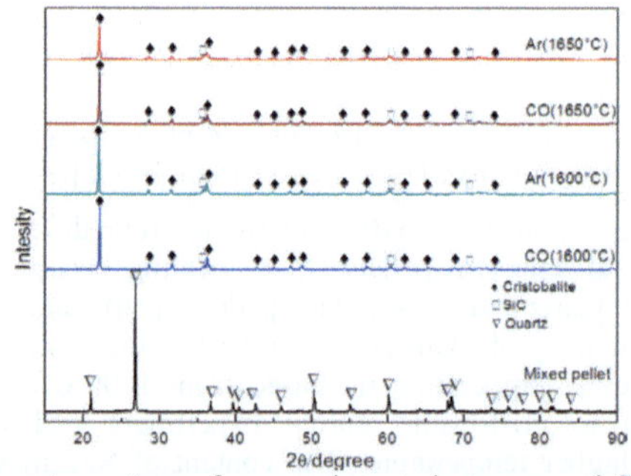

Figure 7 X-ray diffractometer patterns for experiments in argon and CO atmosphere at 1600°C and 1650°C and mixed quartz and carbon black pellets (before heating)

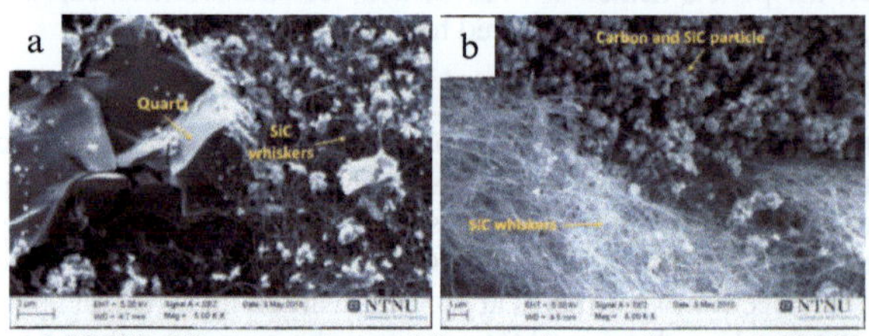

(*a*) mid-section, at 1600°C, CO atmosphere; (*b*) exterior surface, at 1600°C, CO atmosphere;

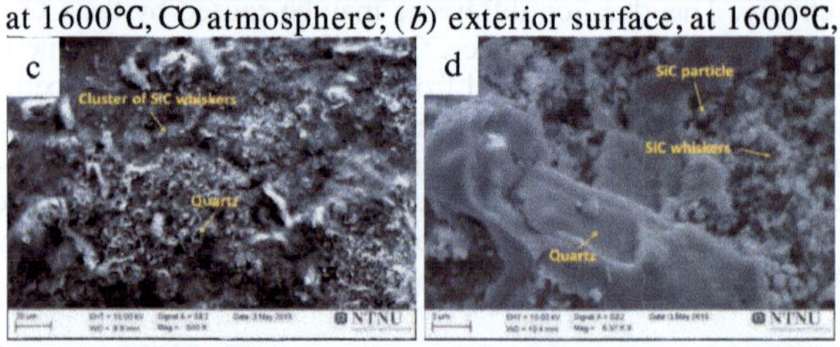

(*c*) exterior surface, at 1650°C, CO atmosphere; (*d*) mid-section, at 1650°C, CO atmosphere;

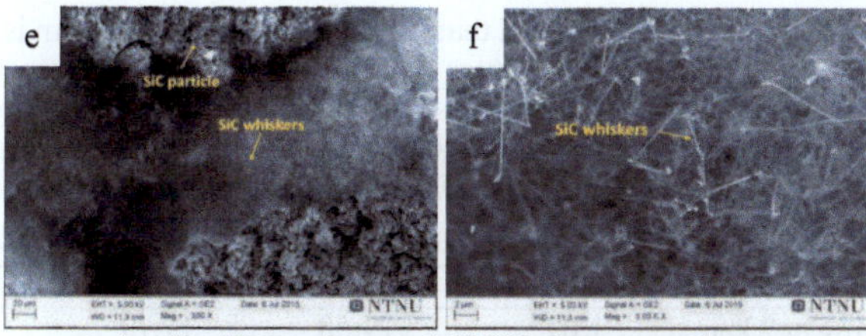

(*e, f*) mid-section, at 1600°C, Ar atmosphere

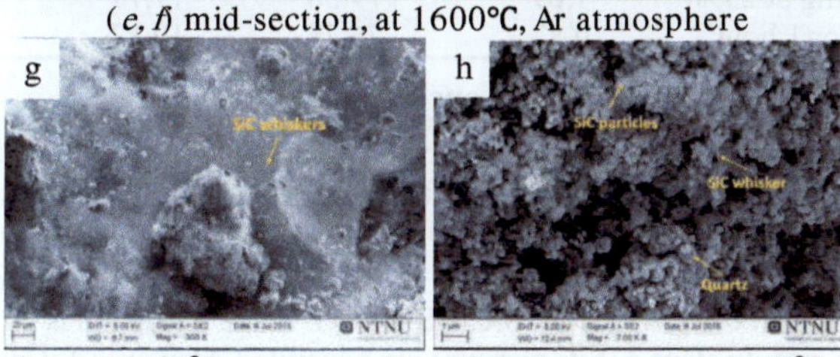

(g) exterior surface, at 1650℃, Ar atmosphere; (h) mid-section, at 1650℃, Ar atmosphere
Figure 8 SEM images for mixed quartz and carbon pellets after heating in argon and CO atmosphere at 1600°C and 1650°C respectively.

XRD patterns in Figure 7 are quite similar with patterns in Figure 5. The quartz is also transformed to cristobalite after heating at 1600°C and 1650°C in argon and CO gas. The quartz and carbon disappeared while the cristobalite and SiC emerged. When comparing the patterns of experiments at 1650°C in argon and CO gas, the intensity of main peaks for cristobalite in argon is lower than CO. This also indicates the higher SiO_2 consumption in argon atmosphere. Carbon is also difficult to find out here.

SEM images of pellet after heating in CO and argon atmosphere at 1600°C and 1650°C are shown in figure 8. The carbon black particles surrounding the quartz grains are not present as it was for green pellets (figure 2b and c). In figure 8a, d and h, the quartz is surrounded by spherical SiC particles according to the EDS analysis. The non-uniformed polyhedron and sharp edges of quartz also disappeared due to the softening of SiO_2 and the reaction on the SiO_2 surface. SiC particles were found inside of the pellets. This due to solid carbon reacting with SiO gas to SiC. In cavities or on the surface of pellets, SiC fibres and whiskers were found. This indicates that the SiC fibers and whiskers are produced from gas-gas reaction between SiO and CO according to reaction 8.

C. Chemical analysis of sample stopped at 23% of weight loss

As mentioned above, when all the carbon is consumed according to reaction 6, the theoretical weight loss is 23%. The reaction at stage 2 can be found when comparing chemical content of pellet at weight loss of 23% and the sample after 1 hour holding time (the weight loss is 26%). One more experiment was conducted with the holding time of 1650°C, but stopped when weight loss reached 23% (held at 1650°C for 29 minutes).The free carbon, SiO_2, SiC and metallic Si in the sample was analyzed. As the data in table 2 indicates, the carbon content is close to 1% for both samples, as the carbon is reacted to SiC in stage 1. If stage 2 is the reaction of SiO_2 and SiC to SiO and CO gas, the amount of SiO_2 and SiC should be quite stable. The content of both SiC and SiO_2 are quite similar, and it is believed that the difference between them is in the area of the variance of the analyses. The chemical analyses is in accordance with the theory of stage 1 being the reaction of SiO_2 and C to SiC and CO, and stage 2 where SiO_2 and SiC is reacting mainly to SiO and CO gas. A minor amount of silicon is also produced according to the result of chemical analyses, but it is not seen in the SEM and XRD.

Table 2 The chemical analyses of sample at weight loss of 23% and 26% (*wt*. %)

	Free carbon	SiO_2	SiC	Metal Si
Stop heating at 23%	1.09	79.58	17.79	1.53
Holding for 1 hour	0.87	76.77	19.17	3.19

Conclusion

The reaction between mixed quartz and carbon pellet at temperatures up to 1675°C in CO and Ar atmosphere can be divided into two stages, the fast stage and slow stage. At stage 1, SiO_2 and C is reacting according to the reaction: $SiO_2(s) + 3\ C(s) = SiC(s) + 2\ CO(g)$, but may occur through subreactions through gas species. The main part of stage 2 is the much slower reaction between SiO_2 and SiC to SiO and CO gas: $2SiO_2+SiC=3SiO+CO$. Stage 2 reaction is quite slow and have a constant reaction rate. In stage 1, SiC particles are generated. However, SiC whiskers are also produced from SiO and CO gas. Higher temperature and argon atmosphere are the positive parameters for SiC formation.

Acknowledgments

The authors acknowledge Elkem and Norwegian Research Council for the financial report through the project "235123 Silicon Production with use of Natural Gas".

References

[1] Schei, A., Tuset, J. K. and Tveit, H., *Production of High Silicon Alloys* (Trondheim: Tapir Forlag, 1998), 25-72.

[2] M. Tangstad, *Metal Production in Norway* (Trondheim: Akademika Forlag, 2013), 97-120.

[3] J. J. Biernacki, "Formation of Silicon Monoxide and Application to the Growth of Vapor-Liquid-Solid Silicon Carbide Whiskers", (Doctoral Dissertation, Fenn College of Engineering, Cleveland State University, Cleveland, Ohio, 1987) , 378.

[4] Agrarwal and U. Pal, "Influence of pellet composition and structure on carbothermic reduction of silica", *Metall Mater Trans B*, 30 (1999), 295-306.

[5] Xiang Li, Guangqing Zhang, Kai Tang,Oleg Ostrovski and Ragnar Tronstad, "Carbothermal Reduction of Quartz in Different Gas Atmospheres", *Metall Mater Trans B*, 46(3), 2015, 1343-1352.

[6] Selvam, N.D.Nair, Paramanand Singh, "Synthesis and characterization of SiC whiskers from coconut shells", *Journal of Materials Science letters,* (17) 1998, 57-60.

[7] Rasit Koc, Sai V Cattamanchi, "Synthesis of beta silicon carbide powders using carbon coated fumed silica". *Journal of Materials Science*. (10)33, 1998, 2537-2549.

[8] Yung-Jen Lin, Chia-Ping Tsang, "The effects of starting precursors on the carbothermal synthesis of SiC powders", *Ceramics International*, 1(29), 2003, 69-75.

[9] Jianfeng Yao, Huanting Wang, Xinyi Zhang, Wei Zhu, Jinping Wei, and Yi-Bing Cheng, "Role of Pores in the Carbothermal Reduction of Carbon–Silica Nanocomposites into Silicon Carbide Nanostructures", *J. Phys. Chem. C*, 111(2), 2007, 636-641.

EPD Congress 2016

ADVANCED CHARACTERIZATION TECHNIQUES FOR QUANTIFYING AND MODELING DEFORMATION

2016 EPD Congress

SYMPOSIUM:
ADVANCED CHARACTERIZATION TECHNIQUES FOR QUANTIFYING AND MODELING DEFORMATION

Session IV

Reviewers:

Khalid Hattar

Marko Knezevic

Wolfgang Pantleon

Rodney McCabe

EPD Congress 2016
Edited by: Antoine Allanore, Laura Bartlett, Cong Wang, Lifeng Zhang, and Jonghyun Lee
TMS (The Minerals, Metals & Materials Society), 2016

Interrupted Quasi-static and Dynamic Tensile Experiments of Fully Annealed 301 Stainless Steel

O.G. Rivera[1], Z. McClelland[2], P. Rivera-Soto[3], W.R. Whittington[4], D. Francis[4], R.D. Moser[2], P.G. Allison[1]

[1]Department of Mechanical Engineering, University of Alabama, Tuscaloosa, AL; Tuscaloosa, AL, 35487, USA
[2]Geotechnical and Structures Laboratory, US Army Engineer Research and Development Center, Vicksburg, MS; Vicksburg, MS, 39180, USA
[3]Department of Mechanical Engineering, University of Puerto Rico, Mayaguez, PR; Mayaguez, PR, 00681, USA
[4]Center for Advanced Vehicular Systems, Mississippi State University, Starkville, MS; Starkville, MS, 39762, USA

Keywords: 301 stainless steel, austenite, interrupted tensile experiment, martensite, phase transformation, split-Hopkinson

Abstract

This research examined the evolving microstructure of quasi-static and dynamically loaded fully annealed metastable 301 austenitic stainless steel (SS). Experiments were performed to an interrupted strain value of 20% and to failure using a tension Kolsky bar (1000/s) and an electromechanical load frame (0.001/s). Electron Backscatter Diffraction (EBSD) identified the microstructural evolution from the as-received condition to the 20% strain level for the high and low rate interrupted samples. This material achieved over 60% elongation to failure with increasing strength as strain rate increased, as expected. Fractography analysis using SEM showed particles in the microstructure and a ductile failure mode. The 301 SS exhibited a greater amount of phase transformation from parent austenite to α′-martensite at the dynamic strain rate when compared to the quasi-static strain rate during the interrupted experiments. This result is indicative of the increased propensity for austenite to α′-martensite phase transformations at the high strain rate.

Introduction

Austenitic alloys with metastable face centered cubic (FCC) microstructures, such as austentic stainless steel of AISI grade 301, are of great interest as a high strength and light weight material, passive safety devices and crashworthiness structure for reducing injury risks to passengers in vehicles [1] and as structural materials [2,3]. The 301 alloy exhibits excellent corrosion resistance, high energy absorption, weldability, high strength, high strain sensitivity, and high ductility when cold worked among other beneficial properties [1,4]. Table I provides the chemical composition of the 301 SS alloy. Based on this chemical composition with low Ni, Mn, and N content (all austenite / FCC stabilizers), austenitic stainless steels such as 301 exhibit low stacking fault energies (SFE) of approximately 16-18 mJ/m^2 [5]. This low SFE portends to the strong propensity of low-alloyed austenitic stainless steels to undergo austenitic to martensitic phase transformations [6]. During a cold deformation process, this material can undergo phase transformation induced plasticity (TRIP) of the face center cubic austenite; first to hexagonal closed packed ε-martensite followed by body center cubic α'-martensite [7]. This phase transformation under plastic deformation is one such intrinsic energy dissipation mechanism that is very interesting for a variety of applications requiring high strain hardening ductility with limited strain rate sensitivity.

Table I: Chemical composition of 301 SS

Carbon	Chromium	Nickel	Silicon	Manganese	Molybdenum
Max 0.12 wt%	16-18 wt%	6.5-0 wt%	<1.5 wt%	<2 wt%	<0.8 wt%

Extensive testing has been performed at quasi-static strain rates to determine flow stress, ductility, hardening behavior of metastable austenitic stainless steels and their propensity for austenite to martensite phase transformations [6,8]. These studies primarily focus on metastable stainless steel grades such as 301, 302, and 304 which have reduced Ni content. The tendancey for austenite to martensite phase transformations reduces significantly in higher alloyed grades such as 316 or nitrogen-charged (e.g., Nitronic) grades due to the increased Ni and N content, respectively. Previously, Hausild et al. [7] performed uniaxial tensile experiments at ambient temperatures on autenitic stainless steels with strain rates from 0.00005 s^{-1} to 0.05 s^{-1} and reported 912 to 866 MPa ultimate tensile strengths and 20 to 31% elongation-to-failure, respectively, with the material giving a strong strain rate dependence. The researchers observed a martensitic phase transformation after low amounts of plastic deformation.

Previous research by Lee and Lin [9] researched 304 SS in compression at different strain rates varying from 800, 2300 and 4800 s^{-1} evidenced that with higher strain rates, higher yield strength and martensite transformations occur. Research conducted by Talonen et al. [10] compared the effects of strain rates in austentic stainless steel (i.e., AISI 301LN and AISI 304) on austenite to martensite phase transformations. The strain rates studied included 3 x 10^{-4}, 10^{-2}, 10^{-1}, 1, and 200 s^{-1}, with their results also confirming that at higher strain rates, higher yield strength and martensite transformation was observed. Additional research by Lee and Lin [9] revealed that at lower strain rates, the material achieved higher strain-to-failure that is anticipated based on a reduction in yield strength and the rapid onset of martensite formation.

Additional research has studied austenitic stainless steels including varying the rate and percent reduction in area by cold rolling and its influence on grain elongation as well as annealing to induce a martensite to austenite reversion for achieving nano- / sub-micrometer grain size to improve material properties [1–3]. In order to refine the grain size, Karimi et al. [2] cold rolled a 301 stainless steel to 90% reduction in area with the reversion of martensite to austenite induced by isothermal annealing at temperatures between 600 and 900°C with varying annealing durations. Also, Misra et al. [3] using a 301 SS that had 62% cold rolled subjected to a reversion annealing of 800°C for 1 s, obtained a near complete martensite to austenite transformation with 850 MPa of yield strength and 45% of ductility. The feasibility of reversion processing and the ability to reduce grain size is promising for production of a variety of components such as thin sheets, wire, and fibers than can exhibit TRIP deformation mechanisms.

In this study, we expand on previous work by focusing on understanding austenitic to martensitic phase transformations in a metastable austentic stainless steel grade AISI 301 tested at quasi-static and dynamic strain rates. Tensile testing was performed at ambient temperature at quasi static and dynamic (1000/s) strain rates including characterization of specimens subjected to interrupted testing at 20% strain as well as experiments peformed to fracture of specimens. Post-test characterization included metallography, fractrography, crystallographic characterization using electron-backscattered diffraction (EBSD), and microhardness measurements. The goal of the research was to identify and quantify austenite to martensite phase transformations in low SFE austenitic stainless steels with the goal of their intregration into protective materials with high energy dissipation.

Materials and Methods

Microstructure characterization and tensile properties of the 301 SS material was evaluated in both the longitudinal and transverse rolling direction. A water-jet machined all the quasi-static and high rate tensile specimens to the same rectangular dog-bone geometry with a 2 mm x 2 mm cross section and a 4 mm gauge length.

Step-wise polishing for grain size and EBSD texture analysis consisted of 600 and 1200 SiC paper followed by Stuers Mol, Nap, and Chem cloths with 6 μm diamond suspension, 3 μm diamond suspension, and Struers OP-S, respectively. As a final step, an Allied High Tech vibratory polisher prepared the EBSD specimens using a cloth with 0.05 μm alumina suspension for 12 hours.

After machining, an Instron 5280 load-frame equipped with a 100 kN load cell performed ambient temperature quasi-static tensile experiments at a strain rate of 0.001/s to either fracture or an interrupted strain level of 20%. The load-frame conducted the tensile experiments in both the transverse and longitudinal plate directions to determine the mechanical behavior and identify any anisotropy in the material.

A direct tension Kolsky bar conducted the high strain rate ambient temperature experiments at a strain rate of 1000/s to either fracture or an interrupted strain value of 20%. The tension Kolsky consisted of: pulley pistons, an incident bar and a transmitted bar, where the specimen is placed in the grip of each bar as seen in Figure 1 and described previously [11–13] by the researchers. The pulley is used to produce tension in the incident bar while the grip maintains the other end of the bar at rest. Next, the bar is released from the grip, distributing the tension along the bar. A high-speed camera correlates fracture morphology and strain data with the strain gauges on the incident and transmitted bars that are coupled with a data acquisition system to resolve the elastic waves, which provide the stress, strain, and strain rate.

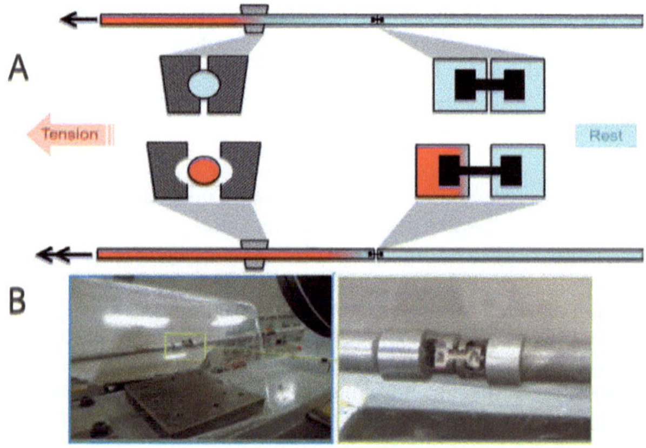

Figure 1. A) Schematic of Split-Hopkinson pressure bar. Red depicts applied tension movement and blue depicts the stationary bar. B) Tension specimen in Split-Hopkinson pressure bar grips.

Fractured specimens of both the quasi-static and high rate specimens provided fractography information by imaging in a FEI Nova NanoSEM 630 Scanning electron microscope (SEM) to identify any differences in fracture morphology observed for the different strain rates. The FEI NanoSEM 630 coupled with a Bruker e-Flash EBSD camera determined the texture analysis of the as-received untested material and interrupted quasi-static and high rate specimens. The EBSD scans obtained the phase maps on a 160x120 μm^2 area with a pixel size of 0.17 μm. The interrupted tensile experiments stopped at a strain value of 20% and coupled with the EBSD analysis evaluated the austenite to martensite transformation for the different strain rates. Interrupted high strain rate SHPB experiments were performed with the use of a connection that served as a fusible link that failed at a stress level corresponding to 20% strain.

Microhardness tests were performed to determine the effect of increased martensite composition on the hardness / strength of the material. Microhardness testing was performed using a Struers Durascan-70 microhardness indenter by indenting the as-received, quasi-static, and high rate samples in multiple locations along the middle of the gage length section of the tensile dogbone specimen.

Results and Discussion

Quasi-static tension tests were performed in both the longitudinal and transverse directions to check for anisotropic stress vs. strain behavior in the as-received 301 SS plate material. The stress-strain response of the 301SS was found to be isotropic in nature with both tested directions yielding similar responses. The 301SS tested at $0.001s^{-1}$ and room temperature exhibited an elongation to failure of approximately 70%, yield strength (YS) of 280 MPa and an ultimate tensile strength (UTS) of 900 MPa. Figure 2 depicts the quasi-static data with an optical image of the fractured specimen. Large amounts of plastic deformation and elongation of the gage section were observed with the typical cup and cone fracture morphology associated with a ductile failure mode.

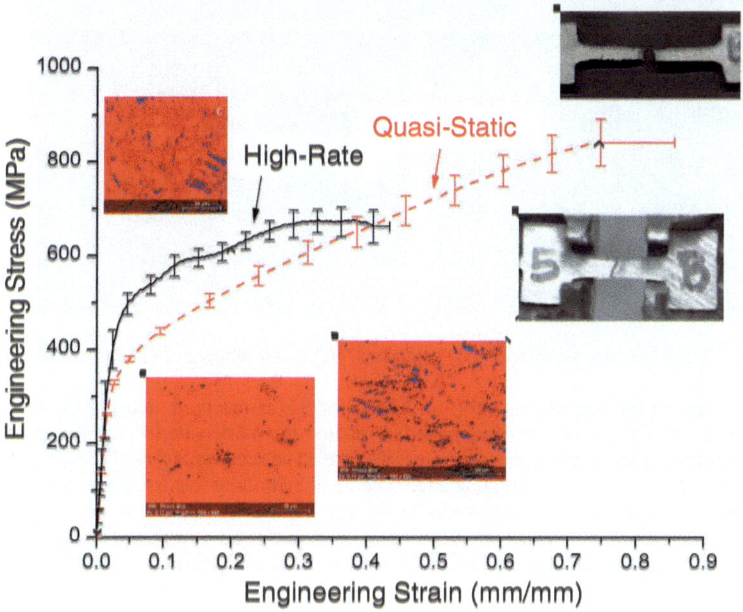

Figure 2. Comparison of the average tensile test between quasi-static and high strain rate. Red line depicts quasi-static and black line depicts high rate. Optical microscope images of the failed dogbones ductile failure for the quasi-static and mixed mode for the high rate.

High rate tensile tests were performed at a strain rate of approximately $1000/s^{-1}$ at room temperature to determine the strain rate sensitivity of the material as well as the effect strain rate has on martensitic phase transformation when compared to the quasi-static (0.001/s) tensile tests. The high rate samples exhibited a lower amount of ductility with an elongation to failure of approximately 40%, but a higher YS of 380 MPa than the quasi-static YS of 280 MPa. Both the yield stress and the stress at the equivalent 20% strain was higher for the high rate specimen. However, the ultimate tensile strength (engineering) was reduced in the specimen tested at the high strain rate. The fractured dogbone in Figure 2 depicts an approximately 45° fracture surface that is indicative of shear bands along with ductile failure mechanisms contributing to fracture. Both the high rate and quasi-static samples displayed very similar strain hardening rates.

Fractography images captured by SEM for samples tested at the quasi-static strain rate evidenced particles in the microstructure and also the ductile failure mode evident by a dimpled fracture surface. Particles can be seen located at the bottom of voids acting as nucleation cites, followed by void growth and finally coalescence. This can be seen in Figure 3B where there are a large number of voids aligned to act as one large inclusion. The surfaces for transverse and longitudinal rolling direction samples did not present any discernable differences in the fractographic surfaces. The presence of particles is anticipated to include various impurities (carbides, sulfides, etc) in the low purity 301 SS studied.

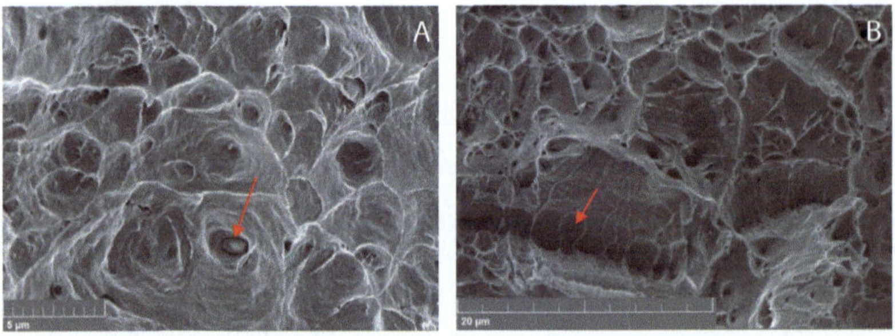

Figure 3. SE SEM fracture surface of the specimens, (A) quasi-static and (B) high strain rate

Figure 4 shows the phase maps collected from EBSD for the as-received, quasi-static, and high rate samples with the FCC austenite (γ) depicted in red and the martensite (α') in blue. Both the quasi-static and high rate samples show long needle like structures representative of martensite produced by plastic deformation. Similar acicular needle like banding formations of martensite in partially transformed austenitic stainless steels have been reported by many other researchers [6]. As would be expected of an annealed stainless steel, the as-received sample did not contain any α'- martensite that was measureable by EBSD. The quasi-static samples microstructure contained 1.9% martensite and that of the high rate sample contained 2.8%. Additional EBSD mapping is needed to further quantify austenite to martensite transformation including studies of the failed specimens adjacent to the fracture surface. It was found that as the strain rate increased, the transformation of austenite to martensite also increased. This behavior was also reported in previous research on 304 SS and is likely associated with the uniform stress state achieved in the Kolsky bar testing as well as the rapid increase in stress throughout the specimen inhibiting localization of plasticity and in conjunction the localization of austenite to martensite phase transformations.

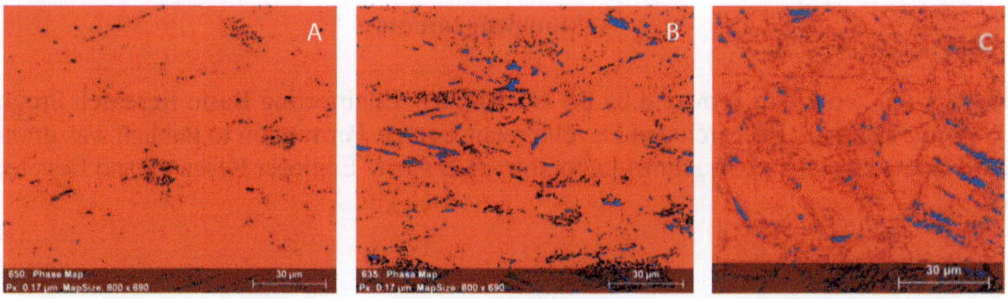

Figure 4. Microstructure phase evolution at the varying strain rates: a) material as received; b) 20% strain at 0.001/s strain rate; c) 20% strain at 1000/s strain rate. Red indicates (γ) austenite and blue indicates (α′) martensite. An increase in martensite with increasing strain rate is observed.

To confirm the microstructural trends observed in EBSD that evidenced increased martensite formation with increased strain rate, microhardness measurements were performed. Hardness is known to increase with martensite content and therefore microhardness measurements can be used to readily assess the amount and spatial distribution of martensite formation. Micro hardness tests revealed the as-received, quasi-static, and high rate samples had hardness values of 185, 277, and 293 HV, respectively. These results confirm those found in EBSD that the specimens tested at high strain rates exhibited increased martensite formation. Additional testing is needed to study the spatial distribution in martensite formation throughout the gage length of the dogbone test specimen. It is hypothesized that martensite formation will be more prevalent along the entire gage length in the high strain rate test specimen given Kolsky bar wave propagation characteristics and the delocalization of plasticity that occurs during high strain rate testing.

Conclusion

Quasi-static and high rate tension tests were conducted to determine the stress strain response and characterize the austenite to martensite transformation in specimens tested to an interrupted strain level of 20% as well as to fracture. The 301 SS specimens exhibited over 70% elongation and approximately 900 MPa tensile strength when tested at a quasi-static strain rate (0.001/!). Fracture was dominated by ductile failure mechanisms. The 301 SS specimens tested at a 1000/s strain rate exhibited an increased yield strength along with a reduction in ductility and ultimate tensile strength. These modifications in stress vs. strain behavior at high strain rates also corresponded with an increase in the formation of strain-induced α′-martensite when compared with the specimens tested at quasi-static strain rates. Relationships between martensite formation and strain rate are a result of internal stress state within specimens subjected to Kolsky testing and the delocalization of plasticity at high strain rates causing a more uniform formation of martensite throughout the specimen gage length. Additional testing to better quantify martensite formation through crystallographic or magnetic properties measurements along with the spatial distribution of martensite measured by microhardness or nanoindentation mapping is needed.

Acknowledgements

Funding for this work was provided by the 6.1 Military Engineering Basic Research Program of the US Army Engineer Research and Development Center. Permission to publish was granted by Director, Geotechnical and Structures Laboratory, US Army Engineer Research and Development Center.

References

[1] M. Eskandari, A. Najafizadeh, A. Kermanpur, M. Karimi, Mater. Des. 30 (2009) 3869–3872.
[2] M. Karimi, a. Najafizadeh, a. Kermanpur, M. Eskandari, Mater. Charact. 60 (2009) 1220–1223.
[3] R.D.K. Misra, B.R. Kumar, M. Somani, P. Karjalainen, Scr. Mater. 59 (2008) 79–82.
[4] AK Steel, Prod. Data Bullletin (n.d.).
[5] R.E. Schramm, R.P. Reed, Metall. Trans. A 6 (1975) 1345–1351.
[6] J. Dash, H.M. Otter, Acta Metall. 11 (1963).
[7] P. Haušild, V. Davydov, J. Drahokoupil, M. Landa, P. Pilvin, Mater. Des. 31 (2010) 1821–1827.
[8] M. Milad, N. Zreiba, F. Elhalouani, C. Baradai, J. Mater. Process. Technol. 203 (2008) 80–85.
[9] W.-S. Lee, C.-F. Lin, Mater. Sci. Eng. A 308 (2001) 124–135.
[10] J. Talonen, H. Hänninen, P. Nenonen, G. Pape, Metall. Mater. Trans. A 36 (2005) 421–432.
[11] W.R. Whittington, a. L. Oppedal, S. Turnage, Y. Hammi, H. Rhee, P.G. Allison, C.K. Crane, M.F. Horstemeyer, Mater. Sci. Eng. A 594 (2014) 82–88.
[12] P.G. Allison, M.F. Horstemeyer, Y. Hammi, H.R. Brown, M.T. Tucker, Y.K. Hwang, Mater. Sci. Eng. A 529 (2011) 335–344.
[13] O.L. Rodriguez, P.G. Allison, W.R. Whittington, D.K. Francis, O.G. Rivera, K. Chou, X. Gong, T.M. Butler, J.F. Burroughs, Mater. Sci. Eng. A 641 (2015) 323–327.

EPD Congress 2016
Edited by: Antoine Allanore, Laura Bartlett, Cong Wang, Lifeng Zhang, and Jonghyun Lee
TMS (The Minerals, Metals & Materials Society), 2016

EVOLUTION OF VOID SHAPE ANISOTROPY IN DEFORMED BCC STEELS

Gregory Gerstein[1], Florian Nürnberger[1], Hans Jürgen Maier[1]

[1]Institut für Werkstoffkunde, Leibniz Universität Hannover, 30823 Garbsen, Germany

Keywords: bcc steel, damage, void development, structure anisotropy

Abstract

Microstructural analysis to support the development of material models provides valuable information to design complex forming operations. Damage models such as e.g. by Gurson consider spherical voids only. However, investigations of bcc steels by scanning electron microscopy reveal that voids exhibit an ellipsoidal shape with progressing deformation. The material's weakening in the Gurson model is taken into account by the void volume fraction. Therefore, a three-dimensional characterization of the void shape offers important data for modelling. Thus, information on the anisotropy of the process of void evolution and accordingly, about the anisotropy of the deformation process is provided. Application of the information regarding the void shape anisotropy within the Gurson model is discussed.

Introduction

All stages of plastic deformation have a different physical-mechanical nature, thus it is required to describe the structure for the corresponding associated models. Usually it is assumed that the formation and evolution of voids in the field of structural defects and inhomogeneity is connected to a substantial development of plastic deformation [1]. In the initial state only a small number of nano- and microvoids are present in steel sheets. Subsequent growth of the voids depends on the size of the local stress and the intensity of the plastic deformation. In the case, where the local stresses in the metal are too small for an intensive growth of voids, the substantial development of plastic deformations can result in the formation of secondary voids. The rate of secondary formation of voids depends on the concentration of the precipitations in the metal, the microstructure of the material, the size of the phase boundaries and the intensity of the development of plastic deformation [2].

For the description of the mechanical material behavior of ductile damage micro-mechanical models can be applied, which are based on the well-known criterion of plastic flow of porous metal of Gurson-Twergaard-Needleman [3, 4]. The physical motivation is given due to void distortion and void interaction with material rotation under shear. Taking into account the experimentally determined critical size of micromechanical damage parameters the initiation of cracks was investigated at the example of compact samples. Therefore the assumed crack position was determined by numerical calculations of thin samples. This model can be used to predict crack initiation.

The relationship of stresses and deformations is determined by Hooke's law and the law of plastic flow which describes the correlation between stresses and the resulting deformation strain for the plastic state taking into account the condition of plastic flow in accordance with the criterion of von Mises. Additional attention is given to the reduction of the cross-section as a result of the formation of damage within the Gurson Twergard model by considering an equivalent void volume concentration. The size of the equivalent void volume fraction f^* is calculated according to Eq. 1:

$$f^* = \begin{cases} f, & f \le f_c, \\ f_c + \dfrac{f_u^* - f_c}{f_f - f_c}(f - f_c), & f > f_c. \end{cases} \tag{1}$$

Here, f denotes the void volume fraction, f_c and f_f denote the critical void volume fraction at incipient coalescence and the void volume fraction at final fracture; f_u^* is defined as the ultimate value of f^* at which the stress carrying capacity vanishes macroscopically as was proposed by Tvergaard and Needleman [4].

Research of the anisotropy of damage and the corresponding void formation can be helpful for a further development of the model.

Experiments and Analytical Methods

For the investigation of damage, that is nano- and microvoids, it is necessary to apply methods which do not influence or modify the near-surface microstructure which formed during the prior plastic deformation. It is intended to ablate a certain layer of the material uniformly. This can be realized by ion slope cutting. Thus, a removal of a metallic layer without damaging the microstructural elements of the material can be achieved by using accelerated argon ions. The thickness of the layer removed is determined by the duration of the ion processing, the voltage at the cathode, the incident angle and the speed of rotation of the sample in the ion slope cutting processing. Different removal depths due to differently oriented grains enable to investigate the particularities of the damage development e.g. at grain boundaries. Here, the void formation and development are investigated as a consequence of plastic deformation in the dual-phase ferritic-martensitic steel DP600. Statistically data of the damage is acquired by taking multiple images: pictures were taken at five locations, lying 50 μm apart. At each location 5 times 5 pictures were taken at a magnification of 5000. In total 125 pictures were analyzed (which complies to the area of 0.5 mm²). The chemical composition of the steel is given in Table 1. For the investigation flat tensile specimens were deformed in a universal testing machine Zwick 100 with optical strain measurement according to the standard DIN EN ISO 6892-1. After deformation, cutting, mechanical grinding and polishing the specimens were subsequently treated by ion slope cutting using an ion-polishing system of the type Met Etch (Gatan Co). The microscopic examination was performed with an in-lens detector in a scanning electron microscope (SEM) of the type Zeiss Supra VP55.

Table 1. Constituent elements of the investigated steel in wt % determined by GDOES

Steel	C	Mn	Cu	Ni	Cr	Mo	Al	S	Si	Fe
DP600	0.10	1.4	-		max. 1.0		0.02-0.06	0.008	0.15	bal.

Results and Discussion

It can be observed that an anisotropic development of voids takes place. Furthermore, the changing of the modulus of elasticity is anisotropically as well. Anisotropy of the damage development is taken into account by applying the Gurson-based Gologanu-Model which considers an elliptical void shape though not the change in the elastic modulus [5]:

$$S_i = \log a_i/a_z \tag{2}$$

with S_i: Gologanu "shape parameter"; a_i: major semi axis along the 0_z direction, a_z: minor semi axis along the 0_x and 0_y directions.

Determination of individual voids includes newly-formed voids, coalescing ones and growing ones. By applying the preparation technique based on ion slope cutting micro and nano voids can be prepared with a high quality [6]. Thus, a detailed three dimensional investigation of voids formed by plastic deformation can be achieved in comparison to established approaches to observe void formation, coalescence and growth on micrometric scale [7]. The shape and size of damage during plastic deformation is determined by the nature and the mechanisms of dislocation slip. The damage development depends on which structural elements are involved in the damage process during plastic deformation. An important role for the damage anisotropy plays the character of dislocation slip inside grains as well as the interaction of moving dislocations with each other and with grain boundaries (Fig. 1a). Frequently, the formation of voids occurs at the boundaries or in the vicinity of grain boundaries thus e.g., large wedge-shaped voids are formed at triple points (Fig. 1b).

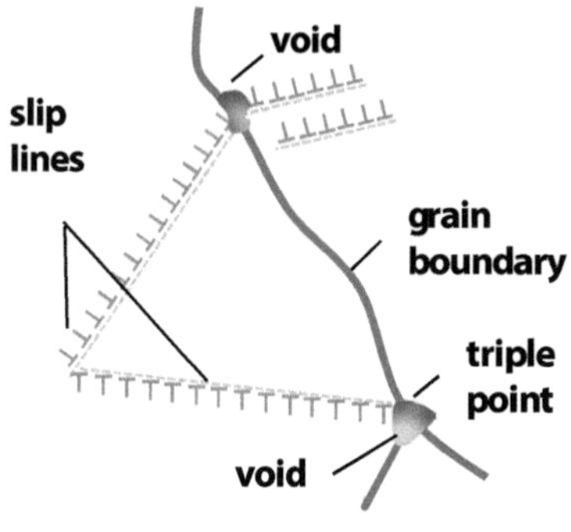

a)

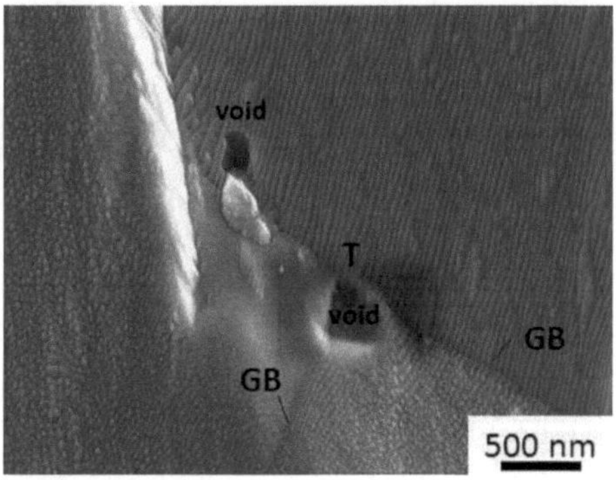

b)

Fig. 1. Damage due to interacting dislocation slip and damage at a triple point, ε = 0.12; a) scheme of interaction between dislocations and elements of the microstructure; b) void formation at grain boundaries (GB) and triple points (T) after plastic deformation.

The three dimensional distribution of voids corresponds to a void development due to sliding in closed packed planes (Fig. 2 a, b, c).

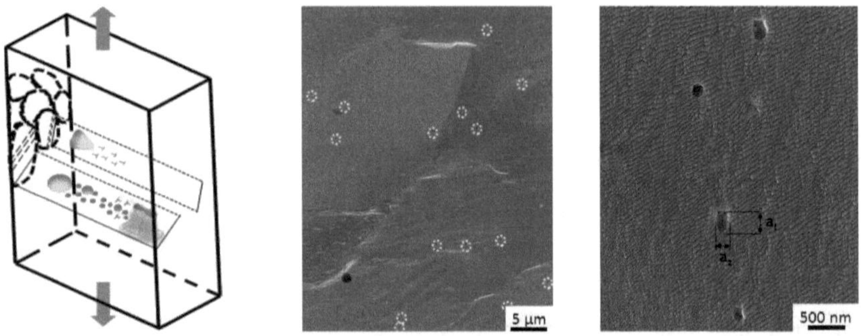

a) b) c)

Fig. 2. Relationship of the dislocation structure after deformation with elements of damage in the volume of the material: a) scheme of interaction between the microstructure elements; b) macro level of the damage: void distribution in the sample after plastic deformation, ε = 0.16 (white dotted circles denote voids); c) micro level of the damage: void distribution in the sample after plastic deformation, ε = 0.16

The anisotropy of the three dimensional distribution of damage is determined by the laws of dislocation slip in close-packed planes of the crystal lattice. This can by illustrated by applying a crystal based geometrical scheme to potential damage in the material volume (Fig. 3a) and by comparing the resulting crystallographic space available for void formation with the presence of real damage in some places in the deformed metal (Fig. 3b)

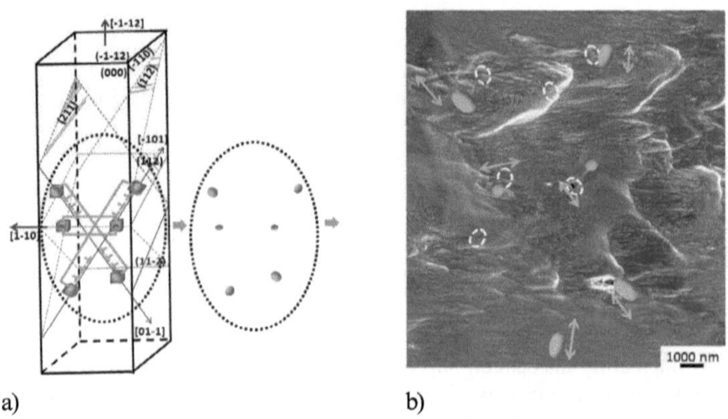

a) b)

Fig. 3. Three-dimensional distribution of voids after plastic deformation, ε = 0.16; development of the cavities due to sliding in the close-packed planes: a) scheme of interaction between the dislocations in closed-packed planes; b) void distribution in the sample after plastic deformation;

the double arrows indicate the direction of a crystallographic allowable displacement of voids produced in a single slip plane, white dotted circles denote voids

Character of dislocations slip and number of involved slip planes is determined by the degree of deformation and defines the shape, size and distribution of voids in the metal volume (Fig. 4a). Thus, when the degree of deformation in bcc steel is less than 0.12, dislocations slip primarily in one close-packed plane {110} is observed, Fig. 4b. Regarding void formation and growth a mechanism of rotation of the matrix material during plastic deformation has to be considered (Fig. 4d and 4c). Figure 4c depicts one possibility of growth and movement of voids by a rotation of the matrix material.

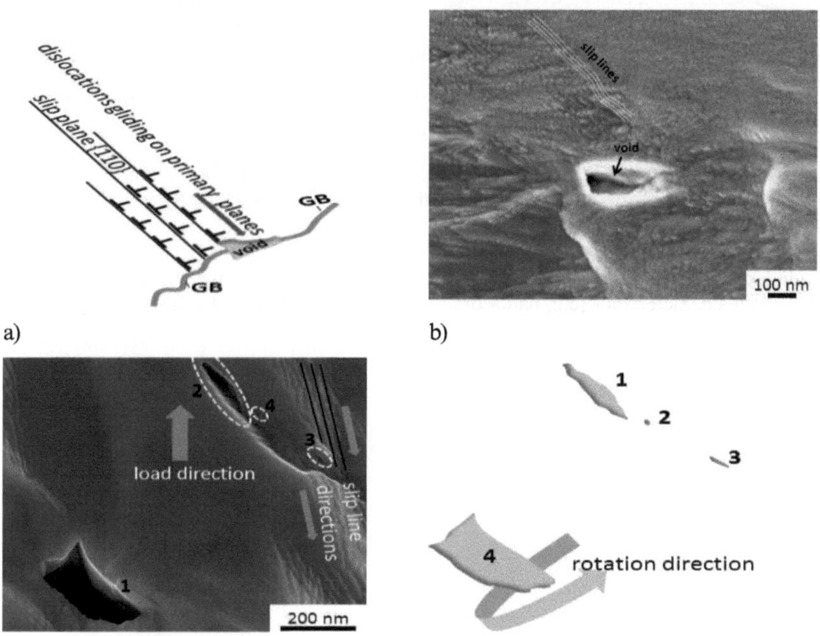

a) b)

c) d)

Fig. 4. Void development in the material volume after plastic deformation, ε = 0.12; a) scheme of void formation according to the mechanism of dislocation slip; b) wedge-shaped void in the material volume; c) void coalescence (1), nucleation (2), growth by dislocation slip (3), growth by rotation (4); d) schematic representation of voids forming and growing by matrix rotation

Naturally, the difference in the mechanisms of void development increases the anisotropy of the void distribution in the bulk material. By increasing the degree of deformation to 0.15 or more dislocation slip is observed in the planes {112} and {123} (Fig. 5a), which is in accordance with [8] regarding the development of plastic deformation in bcc steels. In this case, the growth and development of voids has been accompanied by the development of nano-cracks (Fig. 5b), which requires an adjustment in certain values of the void volume fraction.

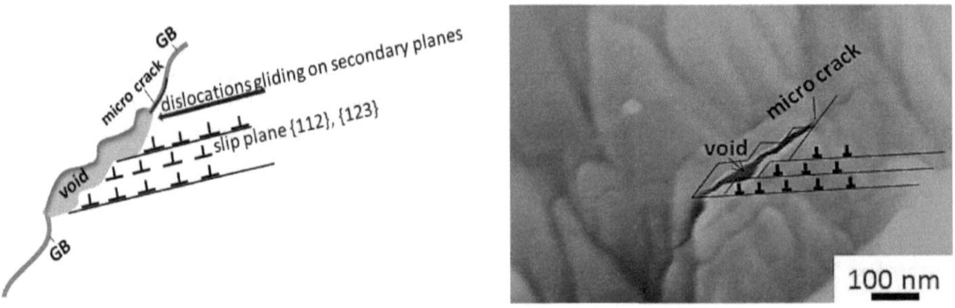

a) b)

Fig. 5. Damage development in the material volume after plastic deformation, ε = 0.23; a) scheme of void and crack formation according to the mechanism of dislocation slip; b) void and micro crack in the material volume

For a detailed analysis values of the void volume fraction *f* must use estimates of the void sizes and other damage characteristics (e.g. nano- or micro cracks) in all three dimensions (Fig. 6).

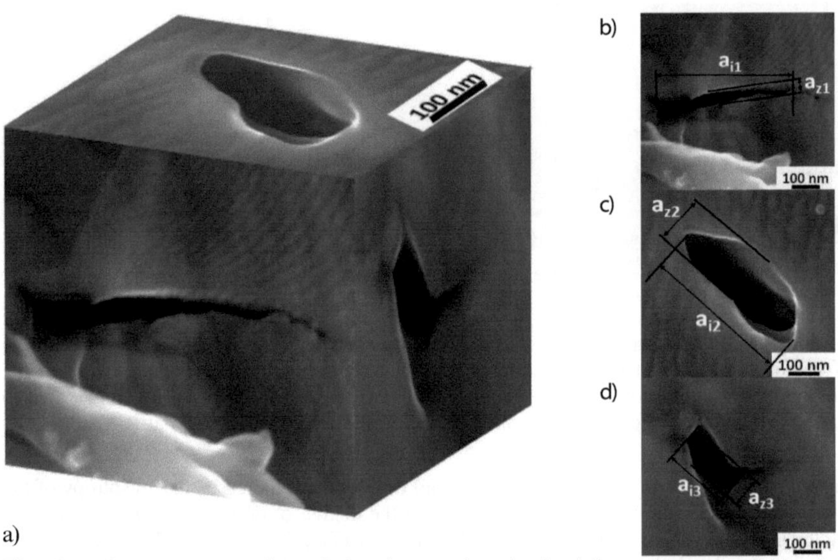

Fig. 6. Volume representation of the damage by plastic deformation, ε = 0.23; a) three-dimensional image of damage; b, c, d) representation of the damage in respective planes

For acquiring statistically significant data of damage by 3D images width ((az2+az3)/2 in Fig.6c and 6d), depth ((ai1 + ai2)/2 in Fig.6a and 6b) and height ((az1+ai3)/2 in Fig.6a and 6d) of the voids were analyzed.
Only by using damage parameters for all three dimensions gives correct information about the three-dimensional void volume fraction.

Conclusions

The exact description of the nature and the anisotropy of the damage distribution in the material volume and changes in the shape of voids with increasing degree of plastic deformation allow calculating the void volume fraction regarding an anisotropic distribution of damage. Changes in the shape of voids in all three dimensions at different degrees of plastic deformation can be taken into account. Using the preparation technology of ion slope cutting specimens could be prepared with a high accuracy regarding the parameters of damage.

Acknowledgment

The authors would like to thank the German Research Foundation (DFG) for their financial support of the transregional collaborative research centre SFB/TR 73 in the subprojects C4.

References

1. Goods S.H., Brown L.M., "The nucleation of cavities by plastic deformation", *Acta Metallurg.*, 27 (1979), 1-16.
2. Argon A.S., "Mechanisms and mechanics of fracture in creeping alloys", *Recent Advances in Creep and Fracture of Engineering Materials and Structures, ed.* B. Wilshire, D.R.J. Owen (U.K. Pineringe Press, Swansea, 1982), 1-52.
3. Gurson, A.L., "Continuum of ductile rupture by void nucleation and growth: Part I – Yield criteria and flow rules for porous ductile media", *J. Eng. Mater. Tech.* 99 (1977), 2–55.
4. Tvergaard, V., Needleman, A., "Analysis of the cup-cone fracture in a round tensile bar", *Acta Metall.* 32 (1984), 157–169.
5. Gologanu, M., Leblond, J.B., Devaux, J., „Approximate models for ductile metals containing nonspherical voids – case of axisymmetric prolate ellipsoidal cavities", *J. Mech. Phys. Solids* 41 (1993), 1723–1754.
6. Radi Z., Havancsák K., Kalácska Sz., Baris A. "Surface polishing and slope cutting by parallel Ar ion beam for high-resolution EBSD measurements" (Paper presented at the Int. Conf. on Nanoscience and Nanotechnology ICONN 2014, Adelaide Convention Centre, South Australia, 2th to 6th February 2014, IEEE Publishing Company), 1-3.
7. Requena G., Maire E., Leguen C., Thuillier S., "Separation of nucleation and growth of voids during tensile deformation of a dual phase steel using synchrotron microtomography", *Materials Science & Engineering* A589 (2014), 242–251.
8. Honeycombe R. W. K., *Plastic Deformation of Metals* (London, Hodder Arnold, 2nd Edition, 1984), 23.

AUTHOR INDEX
EPD Congress 2016

SUBJECT INDEX
EPD Congress 2016

M

N

O

P

Q

R

S

MIX
Papier aus verantwortungsvollen Quellen
Paper from responsible sources
FSC® C105338

If you have any concerns about our products,
you can contact us on
ProductSafety@springernature.com

In case Publisher is established outside the EU,
the EU authorized representative is:
Springer Nature Customer Service Center GmbH
Europaplatz 3, 69115 Heidelberg, Germany

Printed by Libri Plureos GmbH
in Hamburg, Germany